Man and Time

Here and on pages 7, 11, and 15 are 12 paintings representing the months of the year, from the illuminated prayer book *Les Très Riches Heures du Duc de Berry*— perhaps the most celebrated of all the calendar illustrations produced during the Middle Ages. As in most medieval calendars, each month is illustrated by some seasonal activity (usually set against the background of one of the many castles owned by the duke). Though the calendar was begun in 1409 by Pol de Limbourg and his two brothers, it was not completed until many years later by another artist, Jean Colombe. Even so, some of the semicircular records of days above each picture have been left blank. But each semicircle includes the same decorative personification of the sun and the two signs of the Zodiac traditionally assigned to each month of the year.

The first three months of the year: Above, January is depicted by means of the duke and his household at dinner; in the background, a tapestry illustrates a battle scene. A winter landscape represents February, above right: On the left of the snow-covered farmyard, the farmer's wife and servants warm themselves before a fire. Right, March: peasants prepare the field and vineyards for spring near the castle of Lusignan (one of the duke's favorite residences).

Man and Time

J. B. Priestley

Bloomsbury Books
London

© 1964 Aldus Books Limited, London

First published in 1964 by Aldus Books Limited

This edition published 1989 by
Bloomsbury Books an imprint of
Godfrey Cave Associates Limited
42 Bloomsbury Street, London WC1B 3QJ
under license from J. G. Ferguson Publishing Company

ISBN 1870630 67 X

Printed in Yugoslavia

Above, April is represented by young girls picking flowers and lovers exchanging vows against a background of the castle of Dourdan. Above right, a procession of garlanded courtiers accompanied by musicians celebrate the feast traditionally held on May 1. Right, June is depicted by haymaking in fields by the river Seine below the ramparts of medieval Paris.

Editor: Douglas Hill
Art editors: John Wood, Michael Kitson

The author and publisher gratefully acknowledge permission to reprint excerpts from the following works:

Collected Poems by AE, published by Macmillan & Co., reprinted by permission of Diarmuid Russell; *Zapotec* by Helen Auger, copyright 1954 by Helen Auger, reprinted by permission of Doubleday & Company Inc.; *Irrational Man* by William Barrett, © 1958 by William Barrett, reprinted by permission of Doubleday & Company Inc. and William Heinemann Ltd.; "What is Time?" by J. G. Bennett, in *Systematics*, published by the Coombe Springs Press; *Causality and Chance in Modern Physics* by Niels Bohr, published by Routledge & Kegan Paul Ltd., reprinted in the U.S.A. by permission of D. Van Nostrand Company Inc.; *Time and Its Importance in Modern Thought* by F. M. Cleugh, published by Methuen & Co. Ltd.; *Relativity for the Layman* by James A. Coleman, published by the William-Frederick Press; *The Medieval Scene* by G. G. Coulton, published by the Cambridge University Press; "The Evolution of the Physicist's Picture of Nature" by Professor Dirac, published in *Scientific American;* *An Experiment With Time, Intrusions?, Nothing Dies,* and *The Serial Universe* by J. W. Dunne, published by Faber and Faber Ltd. (London), reprinted by permission of Mrs. J. W. Dunne, and *An Experiment with Time* also published by Longanesi (Milan); *Myths, Dreams, and Mysteries* by Mircea Eliade, published by The Harvill Press Ltd. (London), and also by Harper & Row Inc. (New York), Éditions Gallimard (Paris), Compania General Fabril (Buenos Aires), Otto Muller Verlag (Salzburg); *Images and Symbols* by Mircea Eliade, © Harvill Press (London) 1961, also published by Sheed and Ward Inc. (N.Y.), Verlag Otto Walter (Oltan), Taurus (Madrid), and II Saggiatore (Milan); *The Australian Aborigines* by A. P. Elkin, published by Angus & Robertson (London), reprinted by Doubleday & Company Inc. (N.Y.), Einàudi (Milan), Éditions Gallimard (Paris); *Mr. Tompkins in Wonderland* by Professor G. Gamow, published by the Cambridge University Press; "Relationship Between Relativity Theory and Idealistic Philosophy" by K. Gödel, from *Albert Einstein: Philosopher-Scientist* (ed. P. A. Schlipp), published by Harper & Row Inc.; *The Concept of the Positron* by Professor Norwood Russell Hanson, published by Cambridge University Press; *Time, Knowledge and the Nebulae* by M. Johnson, published by Faber & Faber Ltd. (London); *The Tower and the Abyss* by Erich Kahler, published by Jonathan Cape Ltd., excerpt reprinted in U.S.A. by permission of George Braziller Inc.; *The Allegory of Love* by C. S. Lewis, published by The Clarendon Press; *The Uses of the Past* by Herbert Muller, published by Frederick Muller Ltd.; *Living Time* by Maurice Nicoll, published by Vincent Stuart Ltd.; *A New Model of the Universe* and *The Fourth Way* by P. D. Ouspensky, published by Routledge & Kegan Paul Ltd., and reprinted in U.S.A. by Alfred A. Knopf Inc.; "Gnosis and Time" by H. C. Puech in *Man and Time*, papers from the Eranos Yearbooks, published by Routledge & Kegan Paul Ltd. and Rhein Verlag (Zurich) and reprinted in U.S.A. by permission of the Bollingen Foundation; *The Philosophy of the Upanishads* by Dr. S. Radhakrishnan, published by Allen & Unwin Ltd.; *Philosophy of Space and Time* by Hans Reichenbach, © 1958 by Maria Reichenbach, published by Dover Publications Inc. and by Constable & Co. Ltd.; "Precognition and Intervention" by Dr. Louisa E. Rhine in *The Journal of Parapsychology; Dictionary of Philosophy* edited by Dagobert D. Runes, published by Peter Owen—Vision Press; *Time and Eternity* by Professor W. T. Stace, published by Princeton University Press and by Oxford University Press; "Space, Time, and Eternity" by Dr. Gustave Stromberg in the *Journal of the Franklin Institute; Natural Philosophy of Time* by G. J. Whitrow, published by Thomas Nelson & Sons Ltd.; *Rise and Fall of Maya Civilisation* by J. Eric S. Thompson, published by Gollancz Ltd.; *Collected Poems* by W. B. Yeats, published by Macmillan (London), The Macmillan Company (N.Y.), Vallechi Editore (Florence), Éditions Montaigne (Paris), reprinted in German by permission of H. Herlitschka—reprinted by permission of Mrs. Yeats, © 1956.

Above, to illustrate July, corn cutting and sheep shearing in front of the castle of Poitiers (which was reconstructed by the duke). Above right, the castle of Étampes forms the background to August. While harvesters work in the fields or bathe in the river, a procession of ladies and gentlemen set out to hunt with falcons. Right, September—and the castle of Saumur, famous for its wine, for which peasants are shown picking grapes.

Introduction

Most studies of Time have been the work of mathematicians, physicists, or philosophers. Even with their highly technical *expertise*, it has never been easy work for them. And though I have allowed myself some linguistic liberties that many of them would have rejected, this book has not been easy work for me; indeed, I would say it has been the hardest task I ever set myself. This is not because I am not a mathematician, physicist, or philosopher—after all, my approach is quite different—but because the subject itself is at once large and yet peculiarly elusive, like a Moby Dick that may be a specter.

Unlike most writers on Time, who pretend and may even have deluded themselves into believing that they are purely objective, I will confess to prejudice, bias, an approach to some extent directed by feeling. What we have here is a personal essay. It may take note of the attitudes and conclusions of some mathematicians, physicists, and philosophers, but it is not attempting to challenge them on their own ground. And for many of the chapters here, even the professional historian's manner, lofty and impersonal, would have been a fake. But what is more important is that toward the end of this account of Man and Time any attempt at an objective manner would be impossible. The material offered will be itself deeply subjective, belonging to one man's inner world of thoughts, feelings, intuitive ideas, vague impressions, belonging in fact to my own personal encounter with Time. This seems to me the only possible conclusion to such a book. In spite of its size and range, this is then a personal essay. It is a Time-haunted man addressing himself chiefly to all those people he knows from experience to be also Time-haunted.

Readers are asked to note in advance that my Time with the capital initial is the subject of this essay. This is simply for convenience; I am not trying to get away with a little quiet "hypostatization." Time's times, I trust, will not carry this capital letter.

Now I have to thank various people while absolving them from any responsibility for what I have written. I have to thank Miss Carolyn de Sainte Croix, of the Cambridge University Library staff, for material used in Chapter 1, Part One, and in sections of Part Two; Miss Celia Green, director of the Psychophysical Research Unit at Oxford, for material for Chapter 3, Part One, and with her the scientists whose statements I have used; and Mr. Norman MacKenzie and Mr. Douglas Hill of Aldus Books Ltd. for some helpful editorial criticism. Some further indebtedness I have acknowledged in the text.

13

Above, October—sowing on the banks of the Seine near the palace of the Louvre. Above right, to represent November, a swineherd throws a stick into the trees to bring down acorns for his pigs. December, right, is illustrated by a forest clearing near the castle of Vincennes (the duke's birthplace) where, watched by huntsmen, hounds are devouring their quarry at the end of a wild-boar hunt.

Contents

Part One   The Approach

1   The Measurement of Time

During the late Middle Ages, mechanical clocks—then valuable rarities—were often seen as symbols of wisdom and virtue. This painting, from a 15th-century treatise on morality called *The Clock of Wisdom*, depicts a feudal household admiring the splendor of their master's vast, weight-driven clock.

1 The Measurement of Time

In our strange era some of our wealthiest men plot and plan and toil, probably ruining both their health and some of their acquaintances, so that once or twice a year they can, as they say, "get away from it all," going into what is left of the wilderness, often at enormous expense, to hunt game, to fish, to cook on a camp fire—in fact, to live for a while as their remote ancestors lived all the year round. They are trying to escape, if only for a brief spell, from their own time. But I do not suppose they leave their watches at home. They have only to glance at their wrists to know the hours, the minutes, the seconds. The men whose lives they are imitating, at such trouble and expense, knew nothing about hours, minutes, seconds. Sunrise and high noon and sunset would be enough for them, these hunting ancestors of ours, living in family groups or small tribes and for the most part, we imagine, in caves. It is surprising that, as yet, no top travel agency, the kind that organizes expensive safaris, has arranged to let genuine prehistoric cave dwellings to millionaire sportsmen. But what could not be rented out with the caves is their original occupants' notions of time. The new tenants would still be looking, often anxiously, at their watches.

It must have been when their nomadic hunting life was well behind them, when they stopped in one place to cultivate the soil and to rear domestic animals, when they found themselves in larger and larger communities, that men felt the need of being able to tell the time. Sunrise and high noon and sunset would be no longer good enough. Men with fields to till and to harvest, with animals to care for, could not afford to be as negligent about time as were the old hunters, who used up a lot of energy in tracking, pursuing, and killing their prey and who then probably enjoyed lolling around while the women and children collected berries and roots. (Men differ to this day as expenders of energy:

some want to work steadily without ever having great demands made upon them; others feel frustrated if they cannot alternate between a furious expenditure of energy and spells of idling. It is a weakness of modern industry that it offers comparatively few opportunities—and most of them only on a fairly high level—to the second type of temperament, the work-like-hell-and-then-pack-it-up men, responsible for most of our creative ideas and achievements. Men of this type far outnumber the opportunities that modern life can offer them, which is one reason why we are now piling up frustration.)

Men with regular work on their hands, wishing to meet other men to do a little bartering, needed some division of time between sunrise and noon, noon and sunset, if only not to be hanging about idle too long. And, as communities grew, they had their public business to attend to, without too much waste of time. So the desire to divide the day, in a fashion understood by everybody, must have been very strong.

Where there was plenty of clear sunlight (and we know that the early agricultural communities came into existence in subtropical regions), observant men must have known roughly how the hours were going, long before there were any instruments for measuring time. They would notice how the sharp shadow of a rock gradually dwindled as the sun climbed higher, how it extended itself in a new direction as the sun descended and the day wore away. They may even have made marks on the ground. If they had no natural object, like a rock, to provide shadows, they may have soon discovered that an erect stick would serve them even better.

Observation of the movement of the sun and its shadows naturally led to increasingly accurate measurement of both longer and shorter periods of time. Once communities were large, stable, well organized, more or less able to create a civilization of their own, they erected enormous and elaborate structures and edifices that had many purposes, among them the marking out of periods of the year. The Great Pyramid, designed in part for the determination of the equinoxes, was one (and the most impressive) of these structures and edifices; and so, unmatched elsewhere in Europe, was England's Stonehenge, with its relation to Midsummer Day.

It is worth mentioning in passing that these were not ignoble enterprises; they compare favorably with our erection of vast "office blocks"—the cathedrals of a very dubious worship—or our manufacture of bigger and bigger nuclear missiles. They may have demanded forced labor, yet it is not difficult to imagine that helping to build Stonehenge might have seemed

Most early methods of recording time were based on the movements of the sun. Right, two Borneo tribesmen measure the shadow cast by a totem stick. When the shadow at midday begins to shorten, they will start preparing the land for cultivation. Below, the figure carved at the top of the totem stick is a spirit of time.

more life-enhancing than watching television screens and football games appears to be now. And at least one poet—AE (George Russell)—accepted this forced-labor challenge, and had his reply ready:

The whip was cracked in Babylon
That slaves unto the gods might raise
The golden turrets nigh the sun.
Yet beggars from the dust might gaze
Upon the mighty builders' art
And be of proud uplifted heart.

The number of proud uplifted hearts to be found in the canyons between our new office buildings is small, and not increasing.

While these great period-markers were being erected, humbler instruments to tell the hours by sun and shadow came into existence. They were the remote ancestors of what became in the end perhaps the most charming of all our time-keepers—the sundial. But at first there was no dial. The earliest known example of a shadow clock, belonging to Egypt in the second millennium B.C., was T-shaped and had the hours marked along its length. The head of the T was a short bar, a few inches high, and of course at right angles to the marked bar along which it cast its shadow. (This shadow-casting bar, common to all instruments of the sundial type, came

Many of the structures erected by the people of ancient civilizations were concerned with measuring time (among other things). Above, a drawing of a reconstruction of Stonehenge in England, a circle of stones that was probably built between 2000 and 1500 B.C ; below, sunrise at Stonehenge today. Many antiquarians believe that the axis of Stonehenge was intended to coincide with the direction of sunrise at midsummer. Though this correlation must have been only approximate, it would have provided the ancient Britons with a simple way of fixing a particular point of time.

Top, an ancient Egyptian time stick (*c.* 10th century B.C.). The shadow of the crossbar fell horizontally across the stick, which was engraved (as above) with the names of five "hours." At noon the instrument was reversed. The sun's shadow was also used to determine the lengths of seasons : Right, a stone at the Inca city of Machu Picchu (built about 1500), whose shadow indicated the summer and winter solstices.

to be known as the *gnomon*.) It was a weakness of this device that it had to be moved in the morning, to face due east with the gnomon toward the sun, and then had to be reversed, to face west, in the afternoon.

Much later, between the sixth and third centuries B.C., instruments were devised that had not to be turned; and later still a concave surface took the place of the long bar. But the most rewarding development, finally giving us the sundial we all know, came when the gnomon was no longer vertical, pointing to the zenith, but inclined at an angle according to the particular latitude of the place of origin, and pointing to the North Pole. The time could then be told not by the length of the shadow but by its direction; greater accuracy was possible; and so long as the sun shone the dark finger pointed to the time.

We may not be familiar with all the elaborations of this timekeeper (which included a variety intended for the pocket), but most of us have known, and probably regarded with affection, old garden sundials. Yet, however antique and weathered some of them may appear, they do not go as far back in our history as we might imagine. We owe this true sundial to the Arab mathematicians and astronomers of the Middle Ages, and it was a rarity in Western Europe before the end of the 15th century. Mechanical clocks were already being made by that time but it was many years before they ran with any accuracy, so that the sundial was often preferred, and of course entirely succeeded in what we think now is its real home, the garden. I say again, now without any "perhaps," that of all our timekeepers this one has revealed most charm—a charm at once smiling and melancholy. It is this silent clock that has been inscribed with so many mottoes:

Amende to-day and slack not,
Deythe cometh and warneth not,
Tyme passeth and speketh not.

As the long hours do pass away,
So doth the life of man decay.

Give God thy heart, thy service, and thy gold;
The day wears on, and time is waxing old.

Hours are Time's shafts,
and one comes winged with death.

But we will let the most famous sundial motto and my old favorite, the 19th-century British essayist William Hazlitt, together have the last word on such inscriptions:

Horas non numero nisi serenas is the motto of a sundial near Venice. There is a softness and harmony in the words and in the thought unparalleled. Of all conceits it is surely the most classical. "I count only the hours that are serene."

Above right, a sundial (*c.* 1060, from Kirkdale, England) divided into eight daylight "hours." The horizontal line marks sunrise and sunset; the vertical, midday. The gnomon (now lost) originally projected from the junction of the lines. Later sundials were often far more elaborate. The 17th-century brass dial, below, had eight faces, six of which were inscribed with classical mottoes.

Most sundials have some kind of philosophical reflection on the passage of time. Right, the motto on a 17th-century English sundial. Below, the Latin inscription on another English dial of the same period reads: "I shall return, thou never." Equally melancholy is the message in Latin on the 16th-century German dial (below right): "Death is certain, only the hour is uncertain."

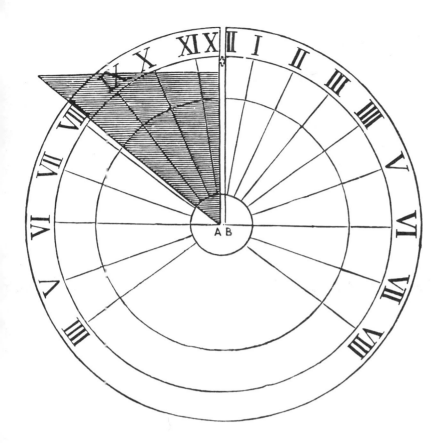

The making of an accurate sundial involves complicated mathematical calculations, since only at the equator will the spaces allotted to the hours be exactly equal. Left, a 17th-century craftsman's working drawing shows the correct hour divisions for a horizontal sundial to be used in London. (The gnomon—whose base would be positioned between the lines marked A and B—is represented on the drawing by the shaded area.)

Left, Holbein's portrait (painted in 1528) of the German astronomer Nicholas Kratzer, who served as "horologer" to King Henry VIII of England. Kratzer is shown working on a polyhedral portable sundial—each of the 10 faces for a different latitude. On the shelf are a cylinder sundial and, for making dials, an instrument incorporating a plumb line. All three instruments reappear among the equipment representing the arts and sciences in the detail, right, from Holbein's portrait *The Two Ambassadors* (1533).

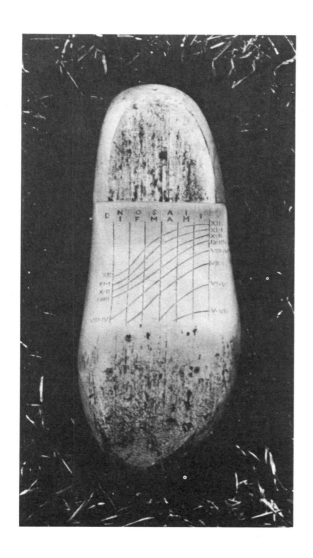

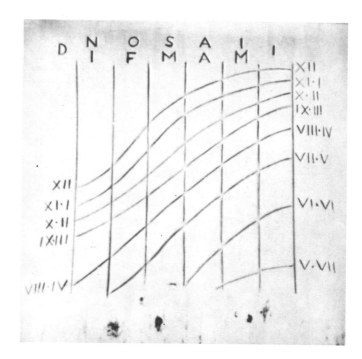

Left, the heel of a clog dial casts a shadow across the scale of curved hour lines divided into months (above). Below, the "human" dial (at Basel, Switzerland) also allows for the varying length of a shadow. If a man stands on the "calendar" of months encircled by 20 stones, his shadow will point to the stone marked with the correct hour.

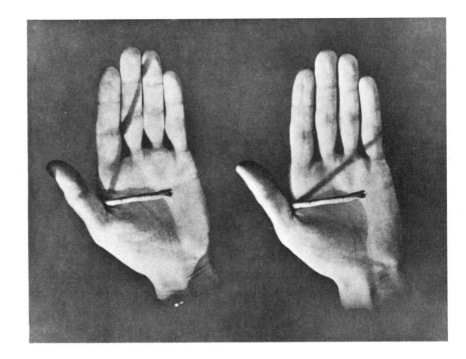

Above and below, two 16th-century German woodcuts depicting "hand-dials"—the countryman's simple version of a portable sundial. Above, a peasant reads the time from the shadow cast on his hand by a stick held under his thumb. Below, a hand with the 12 hours of the day indicated on the fingers.

Telling the time from a hand-dial: The hand is held horizontally, the left hand pointed west for before-noon time and the right hand pointed east for afternoon time. The stick (held by the thumb) is tilted over the palm at an angle approximately equal to the angle of latitude (which is here about 50°, for England).

In the two photographs above the shadow indicates (left) 7 a.m. —the top joint of the third finger —and (right) 10 a.m.—the bottom joint of the little finger. On the right hand (shown below) the time is 4 p.m.—the top joint of the little finger—on the left, and 6 p.m.—the top joint of the second finger—on the right.

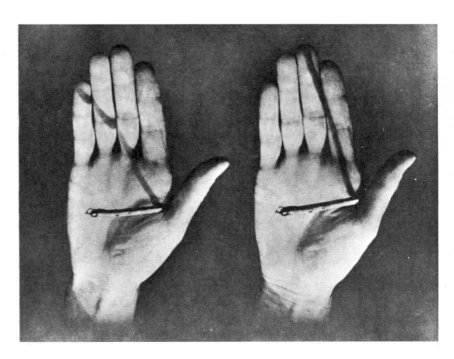

29

A

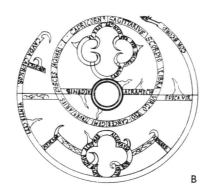

B

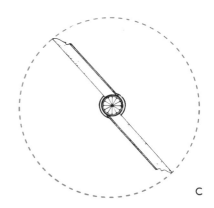

C

Before we leave these sun clocks for other kinds, something must be said about the astrolabe, the most remarkable instrument that the Middle Ages can boast. Its origins go back to ancient Greece; a few astrolabes are still made, for educational purposes; so it may reasonably be regarded as the oldest scientific instrument we know. But it owes most to the astronomical researches and the fine craftsmanship of the Middle East about A.D. 1000. Later it came into regular use in Western Europe, and toward the end of the 14th century Geoffrey Chaucer wrote a *Treatise on the Astrolabe*. As it gave both latitude and time of day, it was used by navigators until about the middle of the 18th century and the arrival of the quadrant.

Using it "to obtain the time" (I am now quoting without a blush from my *Encyclopaedia Britannica*), "first measure the altitude of the sun,

then, having noted the sun's position for the day in the zodiac circle, rotate the rete until the sun's position coincides with a circle on the plate corresponding to the observed altitude. A line drawn through this point of coincidence and the centre of the instrument to a marginal circle of hours shows the time."

If this is not clear then I must refer the reader to the accompanying illustrations of the astrolabe and its various parts, believing as I do that nothing is more tedious than writing or reading elaborate accounts of how ingenious instruments of this kind do their work. But I can add a final note. Because small and easily portable astrolabes were often made, it can be described as the first watch, a sun-and-stars watch—a credit to our indomitable if often misguided species, and a fine legacy from that brief but brilliant Islamic civilization.

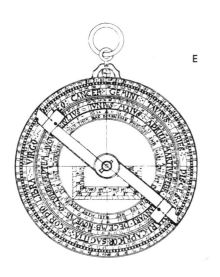

E

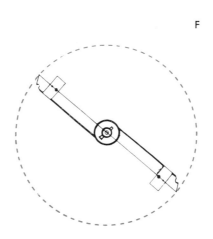

F

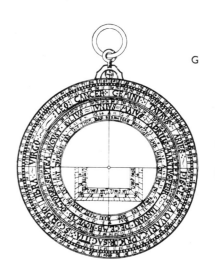

G

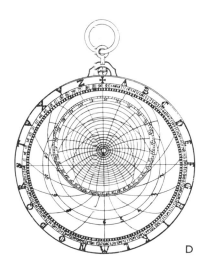

D

The astrolabe was used both for observing the stars and for time-keeping. On facing page and above, drawings by the English poet Geoffrey Chaucer (from his *Treatise on the Astrolabe*, 1391) of an astrolabe from the front (A) and from the back (E); each side is also shown dismantled into its separate parts.

The body of the instrument (D)—called the *mater*—is surrounded by a scale divided into 24 equal hours (here indicated by letters). On top of the mater is a plate engraved with a projection of the celestial sphere. The *rete* (B) forms a map of the brightest stars and is placed over the plate. The outer circle represents the Tropic of Capricorn; the inner, eccentric circle represents the ecliptic—the apparent annual path of the sun.

The *rule* (C) is used for telling the time. First the rete's correct position over the plate is found by measuring star altitudes with the back of the instrument; the rule is then moved until its position over the ecliptic corresponds to the sun's position on that day (found independently). The rule's tip then points to the time on the scale. The rim of the back of the astrolabe (G) is engraved with a scale of degrees; also marked are the signs of the zodiac in relation to the calendar months. The inner circle (showing feast days) surrounds a shadow square, used by surveyors to calculate altitude. The sighting rule (F) is used with the scale of degrees to measure star altitudes.

Above right, a 19th-century British astrolabe; right, three medieval astronomers use an astrolabe to record positions of stars.

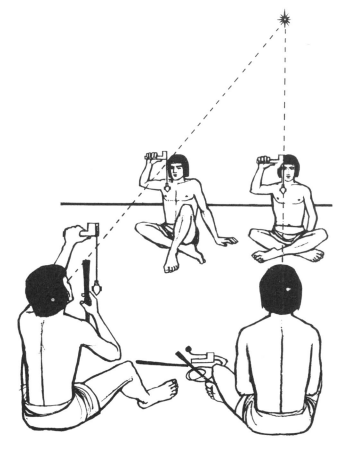

Left, a sighting rod—a simple device used in ancient Egypt to record the transit of the stars. Two of the observers (above), sitting on a line running due north-south, each hold a sighting rod exactly parallel to a plumb line. Through the slit in the top of the sighting rod, they observe the positions of the brightest stars. Such recorded observations provided fairly accurate "time tables" of the night hours. Right, a similar plumb line, held by a shepherd in a 15th-century woodcut, was also used for telling the time in relation to the position of some known star.

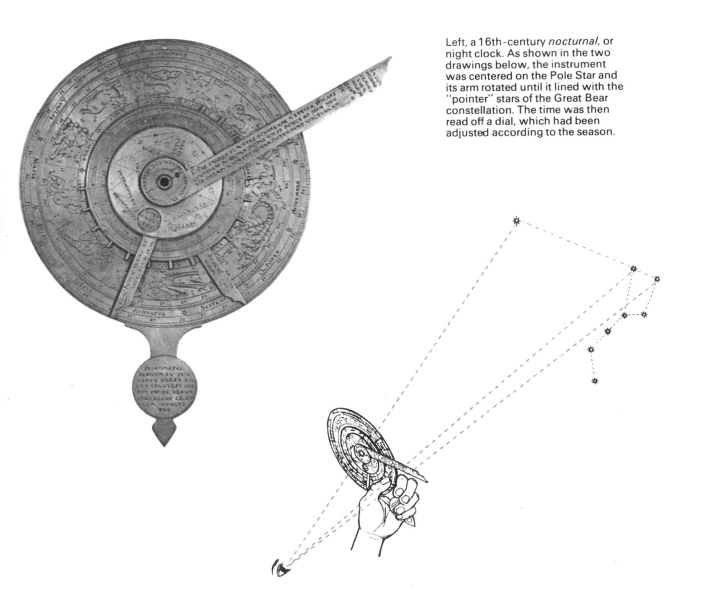

Left, a 16th-century *nocturnal*, or night clock. As shown in the two drawings below, the instrument was centered on the Pole Star and its arm rotated until it lined with the "pointer" stars of the Great Bear constellation. The time was then read off a dial, which had been adjusted according to the season.

Hourglasses (which were probably first used in the 15th century) were a convenient way of measuring some specific length of time, such as a ship's "watch." Above, an ornate set of 18th-century Chinese hourglasses containing powdered marble (from a stonemason's work) that had been boiled in wine. Each of the four glasses marked a quarter-hour. Below, a 17th-century sandglass, used for timing sermons, that today is still fixed to the pulpit of a parish church in Buckinghamshire, England.

There were other methods, however, of measuring time—methods better adapted for places where the sun and clear skies were not often seen, or indeed for nighttime anywhere. One method was burning a marked candle; another was burning oil in a vessel that indicated, however roughly, the hours. In dry countries a method was used that came to have a long history: Sand ran out of one container into another, and with it time ran out. Out of this running sand came finally the hourglass, to give us one of our favorite images of passing time.

The hourglass, with its two pear-shaped glass bulbs, was once a familiar sight on church or chapel pulpits. (Those were the days when sermons were often of an inordinate length, and we know how the hearts of younger members of the congregation would sink when they saw the preacher, who had already used up 60 minutes, reverse the hourglass for the next 60 minutes.) Smaller "sand glasses," as they were often called, have been made to indicate briefer passages of time, and one distant descendant of the ancient desert sand clocks is still in use—the egg-timer. There is a story that I told 30 years ago of a very grim Lancashire widow, who, when asked what she intended to do with her husband's ashes, replied that she was having them put into an egg-timer. "Lazy beggar would never work when he was alive," she added. "He can do summat now he's dead."

It was water, though, not sand, that gave us the aristocrat of these other devices. The water clock or *clepsydra* has had a long history, even though it has never provided us in the West with any of our favorite images of passing time. We associate it chiefly with China, where it came early and was subsequently highly developed; but in fact it originated in Egypt and Babylonia. There were two main types of water clock. In the simpler and probably earlier type, a bowl or other vessel with a small hole in the bottom was filled with water, and it would empty in a given time and possibly, when the water level was compared with marks on the side of the bowl, would show the passage of hours.

The second type, which came to be more elaborately developed, was based on the inflow, the time taken to fill a vessel. A float, rising with the level of water, would carry an arm or pointer that moved over a scale of hours. Later there were further elaborations and complications of this type of water clock, with a variety of mechanical operations being performed, all of them easier and more amusing to illustrate than to describe. This Oriental emperor of timekeepers was probably the most delightful of them all—and was also the most accurate until mechanical clockwork, after some centuries of experiment, finally achieved precision.

Water clocks—more accurate and more adaptable than hourglasses—were used all over the ancient world. Above right, a model of an Irish water-bowl clock: The bowl gradually sank as water filled it through a small hole in the bottom. In the Chinese clock, below, the water emptied from one container to another, each marked off into divisions of time. Similarly, time could be measured from the sinking level of oil as it was burned by a wick. Right, a 17th-century oil clock from Holland.

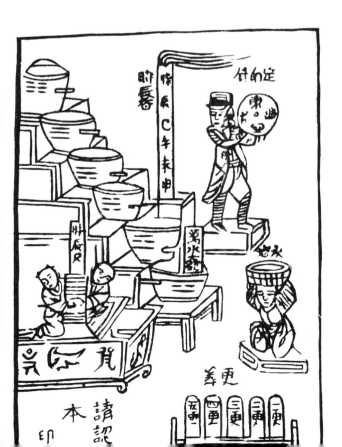

I have known men so fascinated by and infatuated with clockwork that they would spend a good part of their spare time mending, without any charge, other people's watches and clocks. And it is one of them and not I who should now take over this chapter. Never in my life have I wanted to take a clock to pieces and then put it together again. Had I ever been a pawnbroker's assistant, I might have behaved as Charlie Chaplin does in one of the best little scenes in his early films. A man hands over a small metal clock. Charlie looks at its face, then turns it over and uses a tin-opener on the back of it, shakes out the works, examines them and squirts oil at a restless mainspring, clearly decides against this clock, then solemnly removes the owner's hat and sweeps into it the whole mess, finally returning the hat with a slight deprecatory smile. This is the kind of genius that I understand and appreciate, not that which concerns itself with falling weights, mainsprings, and pendulums.

Above, a German engraving of 1698 of a craftsman surrounded by sandglasses. Sandglass making remained an important trade in Europe even after the development of mechanical clocks during the 16th and 17th centuries. Below, the interior of a clockmaker's workshop, about 1580.

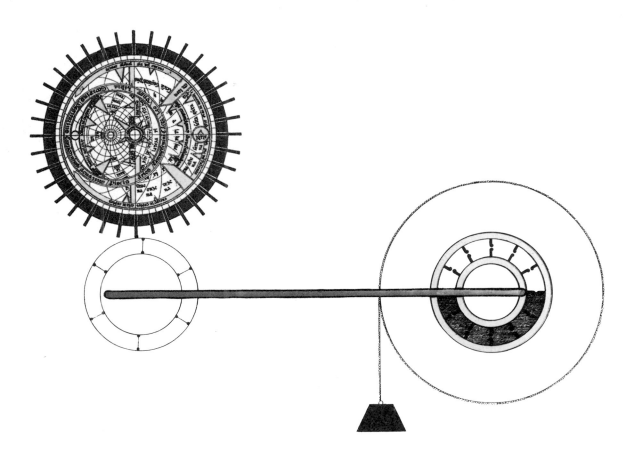

Above, a diagram of a mercury clock (1276). The toothed plate was turned by two cogs linked by a rod—which, though appearing flat, must have looked something like a dumbbell. The instrument was driven by a weight suspended from a cord wound around the right-hand cog, which contained a series of hollow chambers. Mercury leaked through these chambers to check the pull of the weight—a controlling device that is the earliest recorded European example of the ''escapement mechanism,'' forms of which occurred in most later mechanical clocks.

Right, Charlie Chaplin in *The Pawnshop* (1916) as the unmechanical pawnbroker's assistant baffled by the intricacies of a customer's clock.

Mechanical clocks of a sort existed in the 13th century, but the earliest survivors belong to the 14th century, both in England and France. The mechanism of these first clocks was based on a falling weight tied to a rope, which was wound around a revolving drum. They had no dials and probably no means of striking, but they could indicate certain hours so that the keeper of the clock, usually a monk, could sound a bell. Automatic striking came a little later, and monastery clocks struck the seven canonical hours. (It was not unusual in the late Middle Ages for clocks to have faces that showed day-hours and night-hours, with 16 day-hours and eight night-hours in midsummer, eight day-hours and 16 night-hours in winter. This was not new, but was really an inheritance from the distant past, when several ancient peoples, though without mechanical clocks, had made allowances in timing their hours for the changing seasons.) The arrival of the mainspring has no certain date. Leonardo da Vinci made drawings of coiled springs, which may or may not by that time have found their way into spring-driven clocks. If they had not, then the credit for their invention must be given to a Nuremberg locksmith named Peter Henlein, who was making use of them early in the 16th century.

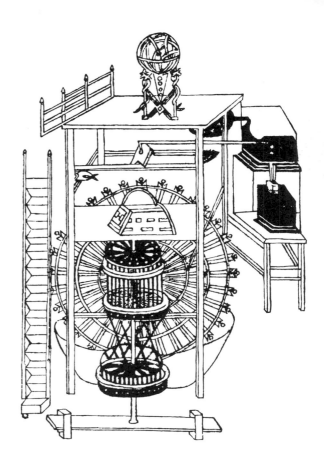

Above right, a drawing of a clock tower from an 11th-century Chinese treatise on astronomy. Inside the tower, an enormous water-powered scoop wheel drove (among other mechanisms) a series of "jacks"—figures that appeared, ringing bells and striking gongs, at specific times of day. The escapement mechanism that checked the wheel's motion—a system of levers that prevented the fall of a scoop until it was full of water—was known in China as early as the eighth century.

Though Galileo discovered the principle of the pendulum in 1581, it was not generally adopted as a method of timekeeping until a century later. Right, a model of the clock incorporating a pendulum and a pin-wheel escapement that was constructed from Galileo's design after his death.

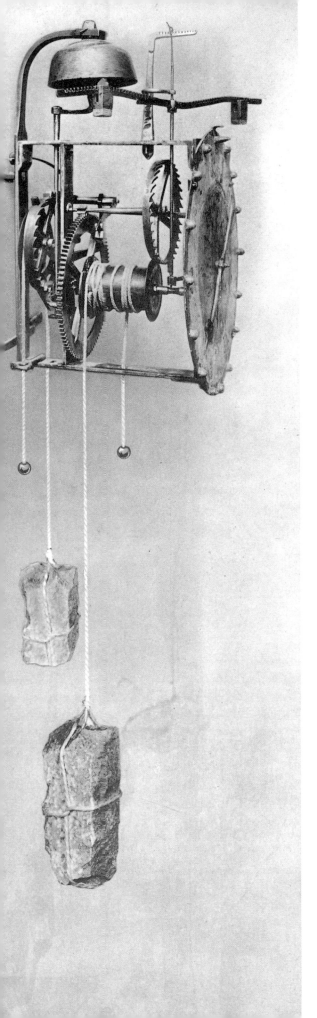

What is certain is that this new driving power, provided by a coiled ribbon of steel, soon made portable timepieces possible. There was not much immediate gain in accuracy, but jewelers' craftsmen enjoyed themselves creating superbly ornamented timepieces, often in the shape of crosses, books, fruit, or flowers. By the middle of the 17th century, when pocket watches were being made, the first hair- or balance-springs began to lighten the mechanism (although these delicate innovations, demanding an increasing skill in workmanship, were not in common use until nearly a century later).

Meanwhile, many cities and towns, especially in Central Europe, had enormous and elaborately mechanized public clocks, which at certain times of day displayed gaily colored figures, often symbolic, that struck the hours. And I have often thought, waiting below for the figures to emerge, that these picturesque old clocks, in which most local citizens still take a pride, represent a meeting point or slight overlapping of two quite different ages, the one to which the symbolic figures really belonged, the other, the beginning of our own age, now capable of highly ingenious mechanism: imagination and faith meeting ingenuity and invention.

The mechanical clock reached its maturity by way of the pendulum, which gave it an accuracy it had not known before. Galileo had understood the principle of the pendulum and was aware of its potentialities, but most of a century had gone before the famous Dutch clock-maker Christian Huygens used falling weights to keep a pendulum going. By the end of the 17th century, new clocks were reasonably accurate; most of them now sprouted minute hands; during the next 50 years or so, the mechanical clock arrived at the summit of its development, both as a timekeeper and as an attractive object among furniture; and

A 14th-century nightwatchman's clock from Nuremberg, Germany, with a dial divided into 16 night hours. Like all early mechanical clocks, the gears (including a bell that rang every hour) were fixed to a drum, which was rotated by the pull of a falling weight. (The second weight acted as a stop.)

since then there have been only minor changes in its construction. To this clock we shall return.

The first patent for an electric clock was filed as long ago as 1840. But none of the early experiments (and there were many of them) was successful. The reliable electric clock and its wide use belong to our present era—which unfortunately is also the era of sudden power cuts that wreck the reliability of its electric clocks. For my part, I have never tried to domesticate this kind of clock, and I used to make its acquaintance chiefly in large establishments like the B.B.C.'s Broadcasting House. There I always found its presence rather disconcerting. Its minute hands had an unpleasant trick of moving in quick jumps, which I often noticed out of the corner of my eye while I was talking or broadcasting, and which always made me feel uneasy. And the domestic type belongs to that kind of all-electric house, common in America, in which everything goes if the current fails: You cannot keep warm, you cannot cook, you cannot have a bath, and you do not even know what time it is.

Left, the vast, ornate face of the 14th-century clock at Wells, England, shows the phases of the moon as well as the time. The clock also included complicated striking devices, such as the figures of two knights who appear each hour above the face. Equally elaborate is the striking water clock, above, from a 13th-century Arabic treatise on mechanics. Among the many decorative figures illustrated are the signs of the zodiac (at the top) and an orchestra of five players.

Left, a rustic scene on an early 19th-century Swiss clock face reflects the idealized view of pastoral life that was fashionable at the time. In contrast, a silver Jacobean death's head watch (above), whose face, right, portrays the Day of Judgment, is a reminder of an earlier, more somber age.

We have today moved, wherever science reigns, from electric to electronic clocks, to which a mere second is an enormous slice of time, ready to be carved into the tiniest imaginable portions. Keeping well away from physical labs, where anything may be happening, I suggest a quick look at our timekeeper-in-chief, the Royal Greenwich Observatory. It still toils on at Greenwich but its larger time department is at Abinger in Surrey. The Observatory uses electronic clocks, which depend on a vibrating crystal of quartz for their timekeeping; there are six at Greenwich, 12 at Abinger. "The relative error of the clocks is recorded to an accuracy of one-hundred-thousandth of a second" (P. Hood, *How Time is Measured*). This error is indicated by a series of continuously moving dials, which do not show the time but the difference in rate between the clocks. The whole set of dials is photographed each day to find the relative error: The real error must be found by astronomical observations. "The quartz clocks have proved, in fact,

One of the many entertaining and ingenious timekeepers of the past was a mechanism that fired a gun at noon. Above, a German meridian sundial with a cannon and lens. The rays of the sun at midday, focused by the lens, detonated the powder in the cannon. Below, a crowd gathered around a similar sundial and cannon at the Palais Royal in Paris in the 1850s.

to be better time-keepers than the earth itself" (Hood). They keep such a steady rate that they can even indicate slight fluctuations (four or five thousandths of a second in a day) in the rotation of the earth.

For all of which we must give thanks to the Astronomer Royal and his staff, though they have probably enjoyed themselves perfecting these arrangements, whereas the rest of us have not felt noticeably happier. And indeed we may feel rather dubious about our quartz clocks that keep better time than the earth itself. What and whose is this time that is better than the earth's? Aren't we her creatures? Isn't she our great Mother? Surely her time, even if it wobbles in thousandths of a second, ought to be good enough for us?

I promised to return, however, to the mechanical clocks that arrived at their maturity about the middle of the 18th century. Though the hourglass and its running sand may have given us the most familiar visual image of passing time,

Above, an English advertisement (about 1910) for a globe encircled by a band marked with the hours. By international agreement, the world is divided into 24 time zones, all of which relate to the time at Greenwich. Below, the "time ball" at the Greenwich Observatory, which is raised between 12.55 and 12.58 a.m. every day and dropped at one o'clock exactly.

it is these clocks that have provided us with the most powerful and disturbing *aural* image. This is certainly true for people of my generation, old enough to have spent at least part of their childhood in the company of Grandfather Clocks. These never seemed mere mechanisms. They were not unlike the mysterious and somewhat awe-inspiring distant relatives or friends of our grandparents who appeared at Christmas or on other special occasions. These elders seemed half human, half mechanical, and so did the clocks.

But we saw and heard far more of the clocks. They cleared their throats, so to speak, before announcing the hours. They seemed to be keeping an eye on us. Their gravely deliberate *tick-tock, tick-tock*, which seemed much louder when we were alone with them, made us wonder what it was that was being *tick-tocked* away, made passing time significantly audible. Moreover, when I was a little boy and my grandmother occasionally looked after me, she would sing after a fashion a popular mid-Victorian music-

hall song (though I cannot believe she learned it in a music hall) about a grandfather clock, which stopped "never to go again, when the old man died." And this symbiotic behavior did not surprise me, just because I felt that these clocks had no mere mechanical existence, that they shared in their tall wooden fashion some kind of life, half-human, perhaps belonging to gnomes and trolls, and so might have developed sympathy and affection for some of their ancient owners, even though their attitude toward me was suspicious and minatory.

Time might seem to spread before me like Genghis Khan's empire, but certain doubts were *ticked*, certain warnings *tocked*, and that throat-clearing before the hour was struck could be ominous. Besides, the grandfather clock I knew best showed me a ship in full sail, before a stiff breeze, but *not going anywhere*. All of which may help to explain why, 60 years later, an elderly writer resolved to forget his lack of qualifications and began to write a book on *Man and Time*.

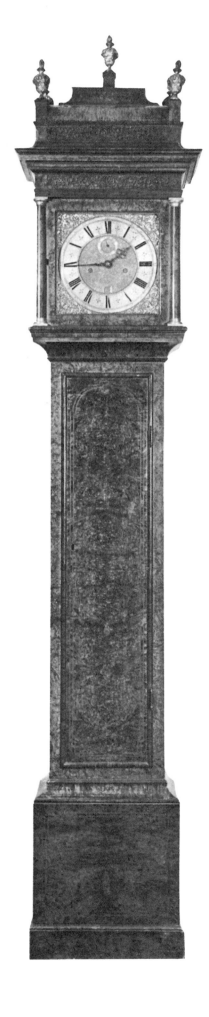

Right, a 17th-century grandfather clock made by England's Thomas Tompion. Grandfather clocks (with the weights and pendulum enclosed in a wooden case) were introduced about 1670, and soon became a traditional feature of English life and a theme for songs and stories. Above, an illustration (from *The Ingoldsby Legends*—a 19th-century collection of popular stories) shows a neglectful husband pursued by the spirit of his wife in the guise of a grandfather clock.

Many grandfather clocks included intricate and sometimes curious details. Left, a "moon" dial from the face of a 19th-century grandfather clock marked off into the 29½ days of a lunar month.

2.

Men knew what month they were in before they knew what time it was. There were calendars of a sort before there were clocks. Early communities in Mesopotamia (a great place for purposeful sky-gazing) had their own calendars, as far back as the third millenium B.C. It was the beginning of a long struggle to make a tidy job out of the rather untidy natural units of time measurement. The creator of the solar system was not keeping man-the-calendar-maker in mind. The Day, one revolution of the earth on its axis, is all right but does not take us very far. The Lunar Month, one revolution of the moon around the earth, seems both reasonable and attractive as a time unit; but its approximate duration of $29\frac{1}{2}$ days is awkward. The Solar Year, one revolution of the earth around the sun, seems more promising still, until we remember its approximate duration, $365\frac{1}{4}$ days, which is plainly asking for trouble. But indeed there was no easy way out for the calendar makers: They were living in one of the untidier corners of the universe.

Sooner or later the calendar men, no doubt to their disgust or grief, had to do some *intercalating*, had to slip in an *intercalary* month; in other words, they had to bring in an extra unit of time

Above, a string calendar from Sumatra represents an elementary method of recording the passage of days in a lunar month. A tally is kept by threading string through each of 30 holes.

To the ancient Egyptians, the first appearance of Sirius in the morning sky heralded the annual flooding of the Nile and the beginning of the fertile season. Right, a relief (*c.* second century B.C.) depicts the cow-headed goddess Isis (who, as Sothis, was identified with the star we call Sirius) watering the land from which corn is sprouting. The bird figure above the corn represents the soul form of the god Osiris.

Since the Egyptians never added a day to their calendar year of 365 days, Sirius's first morning rising occurred on a different date every four years. An extract (left) from a calendar of pharaoh Thutmose III (1501–1447 B.C.) assigns the feast celebrating the star's "going forth," or first appearance, to the "28th day of the third month of summer."

Below, a temple ceiling (c. 100 B.C.) depicts the constellations as personified by the Egyptians. The constellations were divided into 36 decans, each visible for about 10 days. Tables of the stars' positions in each decan were used for calculating the night hours.

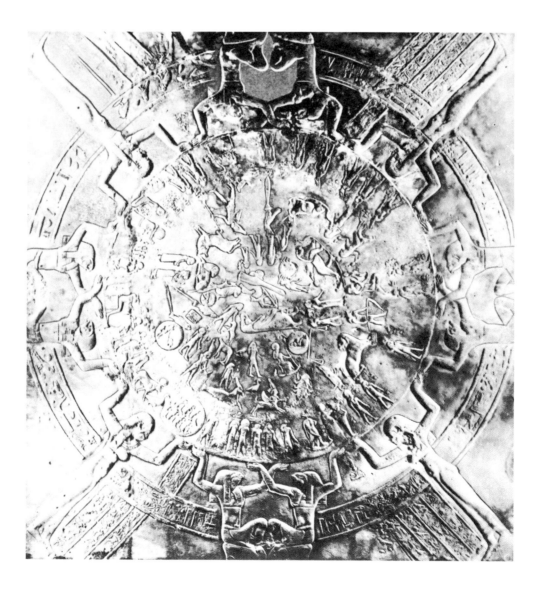

Left, a bronze disk of about the fourth century A.D. inscribed with the initial letters of the Julian Calendar, which was introduced in 46 B.C. This calendar (much the same as the one we use today) remained in force all over the Western world until 1582, when it was slightly amended by Pope Gregory XIII. Below, the meeting called by the pope to inaugurate the Gregorian Calendar, depicted by an artist of the time.

to keep the calendar straight and tidy. But this did not always happen. For example, Mahomet having forbidden any traffic with intercalary months, the old Islamic calendar paraded its 12 moons but found itself with a year of arbitrary length, unable to keep any fixed relation with the seasons. The earlier Mesopotamians had done better than this, not having Mahomet's prejudice against intercalating; they had arrived at a lunar year of 354 days, and after some erratic intercalating, they achieved stability by inserting intercalary months at regular intervals.

The various states in ancient Greece had their own calendars, and the one we know best, the Athenian, also had a lunar year of 354 days, balanced by some irregular intercalation. The Romans originally had a lunar year too, of 355 days; but, in spite of what we may call some intercalatory adjustment, by the time Julius Caesar was ready to take charge of everything and improve it, the Roman civil calendar was three months ahead of the solar year. In 46 B.C. Caesar drastically reformed the calendar, in one step bringing it close to one we can recognize.

It is characteristic of this remarkable man that while he was in Egypt he was equally appreciative of the charms of Cleopatra and the superiority of the Egyptian calendar. (Antony could enjoy Cleopatra, Octavius the calendar; the great all-rounder having gone, specialization was beginning.) The Egyptian civil calendar was unique; it made a real attempt to conform to the length of the solar year; it consisted of 365 days.

Earlier Egyptologists believed that this calendar had been based on the rising of Sothis (Sirius to us), that the year was inaugurated to coincide with the occasion of the latest visible rising of Sothis before dawn, and that, on this assumption, the year 4241 B.C. was the date when the calendar was introduced, making it, as Breasted wrote, the "earliest dated event in human history." But now this astronomical origin of the Egyptian calendar has been repudiated. It is known that there existed over a long period a lunar calendar of 12 or 13 months, and there was plenty of time and opportunity in ancient Egypt to observe that 365 days was the average length of a solar year. And Egypt, unlike

contemporary Mesopotamia, had a centralized administrative system, making it easy to institute a civil year based on the recognition of those 365 days. Feasts and temple services might be regulated by the old lunar calendar, but secular life conformed to the civil calendar.

Julius Caesar began his reform by stretching 46 B.C. until it consisted of 445 days, closing the gap between the civil and solar years. Then he brought in the Egyptian calendar, but with one very useful modification—the extra day every fourth year, our Leap Year. This Julian Calendar, as it is called, long outlasted the Roman Empire. It became the standard astronomical system of reference throughout the Middle Ages. It was still being used by Copernicus in his lunar and planetary tables. Nevertheless, as it aged it gave more and more trouble;

Attempts have been made in the past to ignore the standard method of measuring time. The face of this 15th-century clock at Nuremberg distinguished between summer and winter time. At midsummer, the day consisted of 16 hours, the night, 8—represented by the largest circle of figures. At midwinter, the system was reversed, as can be seen from the numbers nearest to the center of the face.

its year of 365¼ days was about eleven minutes too long; and as the centuries passed they turned hours of error into days. By the 16th century the calendar was well over a week out, and something, it was felt, ought to be done about it.

Nowhere was this feeling stronger than in the Vatican. The Church was worried about the date of Easter, which depends on the appearance of the first full moon after the vernal equinox. The Council of Nicaea, in the year 325, had taken the date of the vernal equinox to be March 21. By the time the Council of Trent met in 1545, the vernal equinox had receded to March 11; and unless something was done, it would eventually coincide with the preceding Christmas. There would have to be a new calendar.

So Pope Gregory XIII inaugurated one, to be known henceforward as the "Gregorian" calendar. Its primary object was to restore the date of the vernal equinox to March 21 and to keep it there. Ten days were annulled, October 5 being called October 15. Three Leap Years were to be omitted in four centuries; which leaves the Gregorian year about 25 seconds longer than the solar year. The calendar year was now to begin on January 1 instead of March 25. And these bold moves brought us the calendar we all know, the one that comes through the post in various shapes and colors every Christmas.

Protestant countries, however, saw no reason why they should submit to Pope Gregory and his Bull. To them the new calendar reeked of Popery and propaganda; there was a Catholic catch in it somewhere. So England, for example, kept to the old Julian calendar until as late as 1752, which meant that 11 days had to be adjusted. A few countries did not change over to the Gregorian calendar until after the First World War. Now it is the one in all our diaries, although the Jewish calendar is still in use. (This is too elaborate to be described here, but, briefly, it combines the solar year and the lunar months, adjusting their difference by intercalating seven months in every 19 years. Even a glance at the enduring intricacies of this calendar helps to explain why so many little Jewish boys grew up to be great mathematicians.)

The division of the day into 24 parts, all of equal length throughout the year, now seems to us inevitable. In the earliest civilizations, however, the hours were not of constant length. The Egyptians divided the daylight day into 12 parts,

A 16th-century German almanac (top and bottom of the page) identifies each month with some traditional activity.

Right, a woodcut of a German calendar (probably also used as a dice board) that was designed as a chart of the most important days in the years 1478–96. The left-hand column lists the "golden" numbers —the number that marks the year's position within a 19-year cycle. This cycle is determined by the phases of the moon, which will recur on the same days of the month every 19 years. Alongside the golden numbers are squares giving the dominical letters (the letters from A to G that relate to the days of the week, starting with A on January 1) and the intervals between feast days.

When Europe adopted the seven-day week, about the fourth century, the days were named after Roman gods. In northern countries (including England), some of the days were later renamed after Nordic gods.

Above left, a "moonface" from a 17th-century Spanish almanac represents Monday, the moon's day. Tuesday was named after the Nordic god Tiw, above (on a fifth-century horn) with the wolf Fenris.

Wednesday was originally the day of Woden—which was the ancient Teutonic name for the Nordic god Odin (the father of the gods), represented above in a decoration from a 10th-century helmet.

Left, a wooden "fan" calendar that belonged to England's archbishop William Laud (1573–1645). The 12 sides of the six leaves correspond to the 12 months of the year. Along the bottom of each leaf, a row of triangles represents the days of the month : The black triangles mark weekdays and the red, Sundays. The various symbols probably indicate Church festivals.

and the darkness into 12 more; but, as the length of daylight increased during the year, the 12 daylight divisions were lengthened. The Babylonians followed a similar practice but divided daylight and darkness each into six parts, thus having a 12-hour day. But even by the fourth century B.C. the Chinese had established a system of 12 equal "double-hours," whereas the Japanese continued to use variable hours until the 19th century. Early Hebrew day-division only recognized six periods, three light and three dark. And the "day" itself began at different times among different peoples.

The "week" was always a purely arbitrary time division (except perhaps among the Jews with their religious observance of one day in seven). Other peoples simply found it convenient to have some time-period between the day and the month: The Greeks divided the month into three 10-day periods, and the Romans had an eight-day week, between the *nundinae* or market days. Our seven-day week owes something to the Jews and probably also to the pagan "planetary week," in which the Sun, Moon, Mars, Mercury, Jupiter, Venus, and Saturn controlled the days in turn. Like our division of the day, this seven-day week now seems inevitable.

It has been challenged, however—notably in Soviet Russia, where an attempt was once made to ignore the week in favor of a system of so many work days and rest days. But the week refused to be banished. I seem to remember H. G. Wells once expressing his discontent with it. He argued that if our week were longer—say, 10 or

Far left, a helmet decoration shows the Nordic god Thor, whose day is Thursday. Above left, a Nordic goddess (c. 700 B.C.)—possibly a predecessor of Frigga, the goddess of Friday.

Above, a 16th-century German woodcut of the Roman god Saturn devouring his children. Saturday— Saturn's day—is the only English weekday whose classical name has not been replaced by a Nordic one.

Above right, a personification of the sun, whose day is (naturally) Sunday. This drawing comes from the same almanac as Monday's "moonface".

Below, a medieval French woodcut of a minstrel entertaining three nuns in the bath. In the days when bathing was infrequent, a weekly bath was often a regular Saturday ritual—a custom commemorated in the Scandinavian name for Saturday, which means "bathday."

Two 16th-century Flemish paintings typify the scenes that were chosen to depict the months in calendars of the time. The winter occupations of woodcutting and sledding represent January (right), while May (far right) is illustrated by the pastimes of boating, music making, and archery.

Right, a 15th-century French calendar for the first 15 days of February includes an illustration of the zodiac sign Pisces. Two columns on the left list the golden numbers and the dominical letters (Sunday is represented by an illuminated letter A). Alongside the columns are the names of the various saints' days. Far right, an 18th-century Swedish wooden calendar. Here, the pictorial symbols representing religious feasts are shown above symbols representing the zodiac signs.

11 days including a three- or four-day week end —a lot of us would work better and play better. This might not please people who have disagreeable or monotonous work to do (and unfortunately such people are still in the majority), but it is probably true that most persons whose work is important to them, persons who are not merely earning a living but bring a certain creative zest to their work, do feel rather cramped between a slow-starting Monday and a quick-ending Friday. They would be happier, too, after a longer and more rewarding period of application to the job on hand, to have an ampler week end in which to relax and play. And some of us, able to decide for ourselves how and when we work or play, do in fact often try to ignore the official week. But only a revolution—and one that would drastically reduce the amount of disagreeable or very monotonous work that now has to be done—could successfully introduce a new and longer week, a genuine week that would bring three more named days between Sunday and Monday or Friday and Saturday.

The earliest system of naming the years comes to us, with so much else, from ancient Egypt, where years were named by the biennial "count," an inventory of taxable property inaugurated in the first or second year of each pharaoh's reign. So men could speak of "the year of the nth occasion of the count" or "the year after the nth occasion of the count." The numbering of individual years of the pharaoh's reigns is thought not to have begun until the end of the Old Kingdom, almost at the close of the third millennium. From Assyria, from about 893 to the mid-seventh century B.C., there have survived the "limmu-lists"—the lists of officials who gave their name to the year. This was probably the earliest really reliable system.

As for the eras, the earliest surviving mention of one occurs in the 13th century B.C. on a monument in Egypt of Ramses II: It is dated in the year 400 of an era begun about 1720 B.C. But the first one to be widely used was the Seleucid Era, which dated from the year 312 B.C. when Seleucid captured Babylon and founded the Seleucid Empire. From the fourth century B.C. onward, the Greeks sometimes used an era that dated from the traditional date of the first Olympiad, 776 B.C. The Romans often fixed their reckoning from 753 B.C., which was the antiquarian Varro's estimate of the date of the foundation of Rome. The Christian era made a curiously late start, being introduced by the abbot Dionysius Exiguus who died about A.D. 540.

There is one odd era we should not forget. It belongs both to astronomy and astrology. Because the earth is not a perfect sphere and is subject to the gravitational pull of the sun and the moon, its movement about its axis is given a twist, and there is a slow but constant change in the direction of its axis. This creates what is called "the precession of the equinoxes." The complete revolution takes nearly 26,000 years (during which the summers and winters of the northern and southern hemispheres will have been reversed), and this vast era is known as the Platonic or Great Year. Astronomy turns into astrology when we divide the Great Year into zodiacal "months," so to speak, and the precession of the equinoxes brings us under the influence of each of the signs of the zodiac. Astrologically, since about 100 B.C. we have been in the Pisces Era (one of the earliest symbols of

Christ was a fish) but are nearing the end of it. By the next century we shall be dominated by Aquarius, bringing a different set of influences— and not before we need them. At present, however, we are not only living in the 1960s of the Christian Era but also in the Pisces Era of the Great Year. And if this adds a touch of wonder and mystery to our lives, then so much the better.

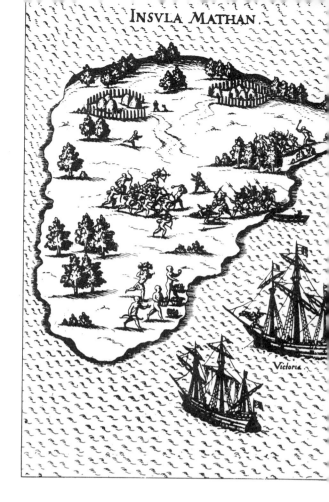

INSVLA MATHAN.

Victoria

3.

If we travel east we go to meet the sun and might find it rising when our watches announced midnight. If we go west then it might still be dark when our time had reached high noon. Fast air travel compels us to notice these time antics. I remember one flight more or less across the central Pacific when in an unchanging clear morning light I was served what appeared to be a series of breakfasts; and a later flight in North Pacific darkness when it always seemed to be time for another late supper. And that was before the big jets began to double the speed, landing us anywhere in a few hours but leaving us feeling uneasier still about time and space and ourselves. (I believe that politicians, foreign-affairs experts, and international relations are the worst victims of air travel.) It is fortunate for us that, 80 years ago, a sensible plan for the world's time variations was agreed upon and put into operation. If we had still to face the appallingly confused variations there were in local times before then, we might now be going out of our minds.

When it is noon at Greenwich, it is probably 12.3 at Burnham-on-Crouch and 11.54 at Salisbury; but this does not help anybody. The only sensible thing to do was to spread the same time out over a fairly wide area, and then, after crossing an agreed boundary, either go forward or backward an hour. So in 1884 the Washington Meridian Conference (for which America was largely responsible) divided the earth into 24 time zones, each zone comprising 15° of longitude. Within the zones there were to be no time variations. Greenwich, which well deserved the honor, was established as the prime meridian.

But somewhere on the other side of the world (to be precise, on 180° longitude) two men might come face to face, and one of them traveling east

Above left, the death (in 1521) of the Portuguese navigator Ferdinand Magellan as seen by an artist of the time. Magellan was killed by natives on the Pacific island of Mactan while he was commanding the first expedition around the world. The survivors of the expedition, on their return home, did penance for mistakenly celebrating various feast days on the wrong dates. In a similar way, Phileas Fogg (above), the hero of Jules Verne's *Around the World in Eighty Days* (1873), almost lost his bet by forgetting the "extra" day that is gained or lost by crossing what is today known as the International Date Line.

By the time Britain adopted the Gregorian calendar in 1732, the discrepancy between the old and new calendars amounted to 11 days. Left, a rowdy scene depicted by Hogarth at the time of the change, when rioters—frightened of losing their wages—demanded to be "given back" the 11 days.

would have gained 12 hours while the other traveling west would have lost 12 hours, and there would be a whole day between them. So the International Date Line was agreed upon, running down the middle of the Pacific, following 180° longitude most of the way, but dodging around in the north to give Alaska and the Aleutians the same dating as America, and scooping some of the South Sea Islands into Australia's dating.

Traveling east, when you cross the Date Line you miss a day, to allow for the hours you have been gaining; and going the other way you have two days of the same date because you have been losing so many hours. (This happened to me in the month of October, 1952, when I had two days with the same date, and very dull affairs they were, both of them; whereas I clean lost forever one fascinating October day when anything might have happened yet I never had a sight or smell of it. There are moments even now when I wonder about that lost day.)

Apart from this tricky business in the middle of the Pacific, the division into 24 standard time zones has worked well. The North American continent has five of these zones, each of them allowed to tell its own time. The Soviet Union has no fewer than 11; but Moscow, an arrogant capital, insists upon all Soviet airports and railway stations (though they may be several thousand miles away) showing nothing but Moscow time. Taking a watch about Western Europe used to be a very simple matter, but now that we have not only Summer Time but Double Summer Time, we have complicated even a modest crossing of the English Channel. Indeed, we are not always sure what time it is when we are still at home.

One thing is certain. Although, down the centuries, we have arrived at some fairly convenient arrangements for the hours, the days, the years, we shall go badly wrong (if we have not already done so) if we imagine that we have now tamed and domesticated Time itself. It may be taming us. But before we consider this possibility (and it must be part of our program here) we must try to discover what various sorts of men have thought and felt about Time.

2   Images and Metaphysics of Time

A detail from *The Triumph of Death* by the 16th-century Flemish painter Pieter Brueghel. A skeleton pointing to a clock is multiplied into an army to convey, in conventional Time imagery, the familiar theme of death's inevitable approach.

2 Images and Metaphysics of Time

I.

In one dictionary of quotations I have counted 288 entries under the heading of *Time*; in another, a larger volume, there are no fewer than 329. As we must expect, many of these quotations come from Shakespeare. *Time*, like *rich*, was one of his favorite words. He was himself both a Time-haunted and Time-defying man. He was also a dramatist, speaking in many voices. According to their circumstances and moods, his characters give Time many different and often contradictory meanings. In various places he makes them say what we have all said, only they mostly say it better. The temptation to quote him is strong, but I feel it must be resisted.

Among the other quotees, a large number, mostly moralists, implore us to make the most of Time. Use it, they tell us, before it goes. Careful men like Seneca and Benjamin Franklin beg us to save it, not to squander it. But who is to decide when it is being saved, when apparently squandered? There are dedicated savers of Time who might as well be social insects: Little good and much harm may have come out of their saving. There are apparent squanderers, leading scandalous lives, who have suddenly brightened the glory of our species. But even they in the end have used their time.

The moralists' root idea—that in Time there is opportunity, which sooner or later must be taken if we do not want to live with regret—is a sound one. Even the Oriental mystic, retiring to a cave to forget "the ten thousand things," must use his time to try to free himself from Time. In this matter of using and not wasting Time, the seers and the sages agree with the old-fashioned moralists; their idea of Time may be very different but they still see opportunity in it, to be ignored at our peril. One party might regard the world as a glittering snare, the other as a place to make yourself snug in; but both parties would be equally horrified at the idea, now so popular, of "killing time."

In medieval and Renaissance art, abstract ideas such as love and death—and Time—were often personified. Below, an illustration from a 15th-century edition of Petrarch's *I Trionfi* represents Time enthroned on a triumphal chariot. Below right, a later and more savage portrayal of Time (from an etching made in 1827 by Britain's George Cruikshank) as the relentless devourer of all that is set before him.

In many of these quotations, representing the wisdom of a thousand market places, Time is praised for being a slow but sure healer and for being a sort of wise detective, sooner or later discovering the truth; it is also denounced as a greedy destroyer, an all-devouring monster like Goya's Saturn. Little or no light is thrown on Time by such praise or blame. The writers use Time as a picturesque shorthand term for all manner of different processes. In such quotations *Time* can be turned into something else; for this reason they are no use to us, Time and not something else being our subject here.

But there can hardly be a more elusive subject. Either it recedes or transforms itself on our approach. Perhaps Carlyle, whom few read nowadays, is our man here, simply thundering away with "That great mystery of TIME, were there no other; the illimitable, silent, never-resting thing called Time, rolling, rushing on, swift, silent, like an all-embracing ocean-tide"— and so on. Add to him a few select and more amusing passages from his contemporary Lewis Carroll (the Mad Hatter's tea party, for example) and we might begin creeping a little closer to our subject. To Carlyle Time is a mystery, to Lewis Carroll it is a joke; and both of them do better than their fellow authors, with their snappy judgments and platitudes about Time, pretending to know everything when in fact they know nothing. The idea of Time as a combined joke-mystery is not to be despised.

Our favorite images of Time are of course all these tides, floods, rivers, streams—anything, it appears, that suggests moving water. *The forward-flowing tide of time.* Or, and I think better: *Time's waters will not ebb nor stay.* And perhaps the most familiar of all, from Isaac Watts's famous hymn: *Time, like an ever-rolling stream, bears all its sons away.* This seems a fine image until we begin to think about it (though, to be fair to Watts, he probably never intended us to think about it). These sons that are being borne away—where are they going? Obviously, they are being floated into the past. Some of them were around last year, now they have gone. This seems all right until we realize that, if this stream is bearing all its sons away, it must be flowing from the future into the past. And if so, tomorrow's rough water might drown somebody yesterday.

And if we feel this will not do, and we reverse the movement of the stream, making it roll from the past into the future, then nobody would have been borne away. We would all be floating happily together, catching an occasional glimpse of Alfred the Great, Chaucer, and Nell Gwyn. This is a fine fantastic idea, everybody in the past floating into the future, but it is certainly not what the hymn-writer meant.

Moving water has been our favorite image of Time probably because it does suggest the stealthy but irresistible passing of the days, months, years. Intellectually it is not a satisfying image because oceans and tides have shores, floods and rivers and streams have banks, so that we cannot help wondering what these shores and banks are, what it is that stands still while Time keeps flowing. On the other hand, it is worth remembering that water is one of the most powerful symbols (frequently making its appearance in dreams) of our dark unconscious life. Perhaps Time the destroyer rises out of these depths.

There is more force in our complaints (where the poets can be found crying in chorus) against Time the destroyer than in our compliments to Time the healer or the wise detective. Most people can join in this chorus because sooner or later they feel they are being changed for the worse. They seem to themselves in their latter days to be poor creatures and in a world that is poorer still. ("There's Capt. Burney gone!" Lamb wrote to Wordsworth. "What fun has whist now?" The quotation editors ignore that cry, yet there is behind it a universal lament.) Even Emerson, a comparatively cheerful sage, can declare: "The surest poison is time." We are so haunted by these ravaging and devouring aspects of Time that some recent writers, perhaps after seeing a film run backward, have told us how much happier we would feel if Time could be reversed. Then we should see thieves stealthily

The traditional figure of Old Father Time portrayed by two well-known satirical artists. Above, the 18th-century British artist William Hogarth depicts Time puffing smoke at a painting (already slashed by his scythe) to age it—and therefore make it more valuable. Right, a cartoon (1846) by the French artist Honore Daumier shows Time peering into a cannon in search of the date of the next war.

bringing presents of jewelry to strangers, millions of young men rising out of war graves, and Sunday editions of New York papers being magically transformed into green trees again.

I have no desire to edit a dictionary of quotations—and indeed I have never made a habit of collecting them—but here are a few on Time that seem to me worth reading. They are not all running one way; they may seem, especially at first sight, to contradict one another; but they are alike in cutting across the accepted con-clusions of the pulpit and the market place, in giving our minds at least a little jolt. I offer them without any names attached;

You talk of the scythe of Time, and the tooth of Time: I tell you Time is scytheless and toothless; it is we who gnaw like the worms, we who smite like the scythe.

To things immortal, Time can do no wrong,
And that which never is to die, for ever must be young.

In order to see famous hills and rivers, one must have also predestined luck; unless the appointed time has come, one has no time to see them even though they are situated within a dozen miles.

For everything that exists and not one sigh nor smile nor tear,
One hair nor particle of dust, not one can pass away.

There are optical illusions in time as well as in space.

Eternity both enfoldeth and unfoldeth succession.

Do but suppose a man to know himself, that he comes into this world on no other errand but to arise out of the vanity of time.

Think that you are not yet begotten, think that you are in the womb, that you are young, that you are old, that you are dead, that you are in the world beyond the grave, grasp all that in your thought at once, all times and places.

And the authors, not in the order above, are Blake, Chang Ch'ao, Cowley, some Anon. in *Hermetica*, William Law, Nicolas of Cusa, Proust, and Ruskin.

2.

Time cannot be reduced to mere change. It is true that without change in some form or other, there would be no Time. This has been denied, chiefly because a certain amount of change has been cheated into the picture. If we try to imagine ourselves in a world without sound or movement, with nothing stirring, without even our breathing or heart-beats, we must agree that we cannot have Time there. Time may not be merely something happening, but unless something *is* happening, there cannot be Time.

People who deny this do not completely freeze the scene; they put their living selves in it; and then of course they would be aware of Time just because something would be happening, and change, no matter on how minute a scale, would be at work. With no possibility of anything changing, within or without, Time would vanish. No change, then, no Time.

Change itself, however, does not give us Time. Suppose we found ourselves in a mad world, created by a surrealist demiurge. A sun like ours rises and sets and there is darkness; then three blue suns follow one another across the sky; then there is a lot of darkness, finally dispelled by a colossal double sun that glares and glares at us until we are sick of it; then a twilight into which six multicolored moons arise; and so it goes on, the scene forever changing, without repetition or rhythm. Even in such a world, of course, somebody could say, "We met when there were those three blue suns, remember?" so that some faint notion of time would be struggling through. But it would be a very dim and distorted notion, not our Time at all.

For Time as we know it, we need both change and not-change, some things moving and others apparently keeping still, the stream flowing and its banks motionless. This may seem a too obvious point to make. I do not think it is, chiefly because, in wider fields of speculation, the point has often seemed to me to have been completely missed. One philosopher tells me that all is flux, nothing remaining the same. But how can he know this? If everything is changing, including himself, how can he know that anything is changing? There could be no standard of comparison, no point of reference.

Similarly, if another philosopher tells me that when he examines his mind and finds there nothing fixed, only an endless flicker of thought and feeling, he seems to me to be forgetting that, unless the searching and reporting self is steadier and more reliable than any flicker or flux, his report is useless anyhow.

Time is a River without Banks, painted in the 1930s by Russian-born artist Marc Chagall. In this highly personal interpretation of the familiar concept of Time as a flowing stream, recollections from the artist's childhood—the fish, the violin, and the family clock— are set against the background of the river. On the bank, two lovers probably represent the timeless quality of love.

Everything in the end may change, but unless the changes take place at very different speeds, we could not be aware of any of them. And though I may appear to have done so, I have not in fact wandered away from our subject. With this relation between what is changing and what is not changing, or between changes that are fast and those that are slower, the problem of Time is bound up—as I hope later to demonstrate.

Into this problem, I do not think our inner sense of duration—*psychological time*—takes us far, at least not on the level on which it is usually examined. Of course it is important and must be discussed, for it is the time we really live with. Clock time is our bank manager, tax collector, police inspector; this inner time is our wife.

Notice how inner time behaves, as if it had centuries of chronological time at its disposal. As soon as we make full use of our faculties, commit ourselves heart and soul to anything, live richly and intensely instead of merely existing, our inner time spends our ration of clock time as a drunken sailor his pay. What are hours outside seem minutes inside. Yet no allowance is made for this prodigality; our calendar time is not extended as a reward for living generously. If we wanted to stretch out our allotment of outer time, to make the most of it in a miserly fashion, we would take care to keep ourselves only half-alive, spend yawning hours (they would be like whole days to our inner time) with boring and mechanical people or with dreary little pursuits.

Failing that, we might do what many important and successful men always appear to have done: appoint a kind of policeman from outer time to control the antics and vagaries of our inner time. When one of these important and successful men awards us 15 minutes of his outer time, we can often see that policeman in his eyes, hear him in the guarded voice. But although this control of inner time might bring more success and importance, few of us feel it would enrich experience and make for a good life. These seem to demand a certain recklessness in our relation with our inner time, as if as we move inward we can defy outer time, the clocks and the calendars, like immortal beings. We are not immortal beings, but we should often behave as if we were.

There is of course a marked change in our experience of inner time as we grow older. I never overhear children crying "Oh, Mummy, what shall we *do*?" without being suddenly taken back to those empty immensities of inner time that children know when they are feeling bored, those afternoons that were like slumbering years. I doubt if I ever feel bored now, so long as I am left alone and no attempts are being made to entertain me; and the time between 2 P.M. and 4.30 P.M. goes flashing by, those huge afternoons of childhood having vanished long ago. This means I miss a lot of fun and excitement, but it also means that I no longer know those deserts of boredom, where children find themselves miserably straggling and cry "What can we *do*?"

We are told that we are at the mercy here of our bodily processes. As they slow down with age they persuade us that time outside us is being accelerated. A rebellious liver and two struggling, ailing kidneys are probably responsible for my feeling that afternoons are dwindling and dwindling and that the years are now so short it is hardly worth while learning to date them properly. (I am now in the middle of 1963 and it still looks an odd stranger.) On the other hand, the incorrupted high-speed metabolism of childhood can by unconscious comparison make time seem to crawl, so that Christmas or the seaside holiday will never arrive.

People dealing with children should always remember this different time scale and its magnification of things. When children meet cruelty or even unkindness or neglect, there seems to them no end to it, they face eons of menace and misery and hopeless bewilderment, they feel lost in a nightmare world that no adults, however much they might suffer, ever stumble into again.

Some persons tell us they always know how clock time is going, without consulting their watches. I have never pretended to have this gift, yet on many occasions, suddenly finding myself awake in the dark, I have said to myself, "It is half-past two" or "It is nearly four o'clock," and have turned on the light to discover that my guess, if it was a guess, had been correct. People who are awakened almost every morning by

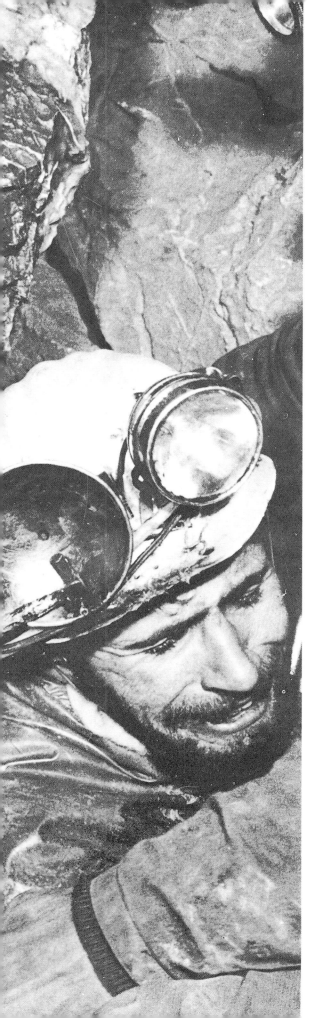

alarm clocks generally allow themselves to sleep until the bell starts ringing. Other people who rarely use alarm clocks and rather shrink from being brassily hammered out of sleep (as I know from experience because I am one of them) somehow contrive to awaken just before the alarm goes off.

Nobody knows how this is done, just as nobody knows how hypnotic subjects, being told to do something at a particular time, perhaps several hours after they have come out of hypnosis, will be punctual in their action to the minute. It seems that something in us below the level of consciousness can if necessary keep us informed about clock time, but without any clocks being involved. A further complication is added when we remember that our unconscious, where this little watchman must live, is not itself committed to clock time, and indeed has ideas about Time quite different from those of consciousness.

To arrive at more confusion let us consider a young French speleologist who was in the news a year or two ago. He undertook to spend some weeks underground alone, far out of sight and sound of his fellow men, and without any means of discovering how time was passing. When at last he was brought up into daylight again, what happened? He found to his astonishment that he had been far *longer* in his cave than he had imagined. We must note first that no intuitive sense of time had worked for him; that little watchman from the unconscious had gone off duty, outraged by this new style of life. Then, as we might expect, his psychological time had lost touch with clock and calendar time. But it had worked the wrong way. Instead of persuading him he had been down there for a fortnight when actually he had been there only a week, a natural exaggeration when he must have been feeling bored and lonely, it went into reverse, making an actual fortnight seem like a week.

The French speleologist Michel Siffre, emerging in September 1962 after two months in a subterranean cave. Deprived of any means of telling the time, Siffre found that he had completely misjudged the length of his stay underground, which was far longer than his "inner clock" had estimated.

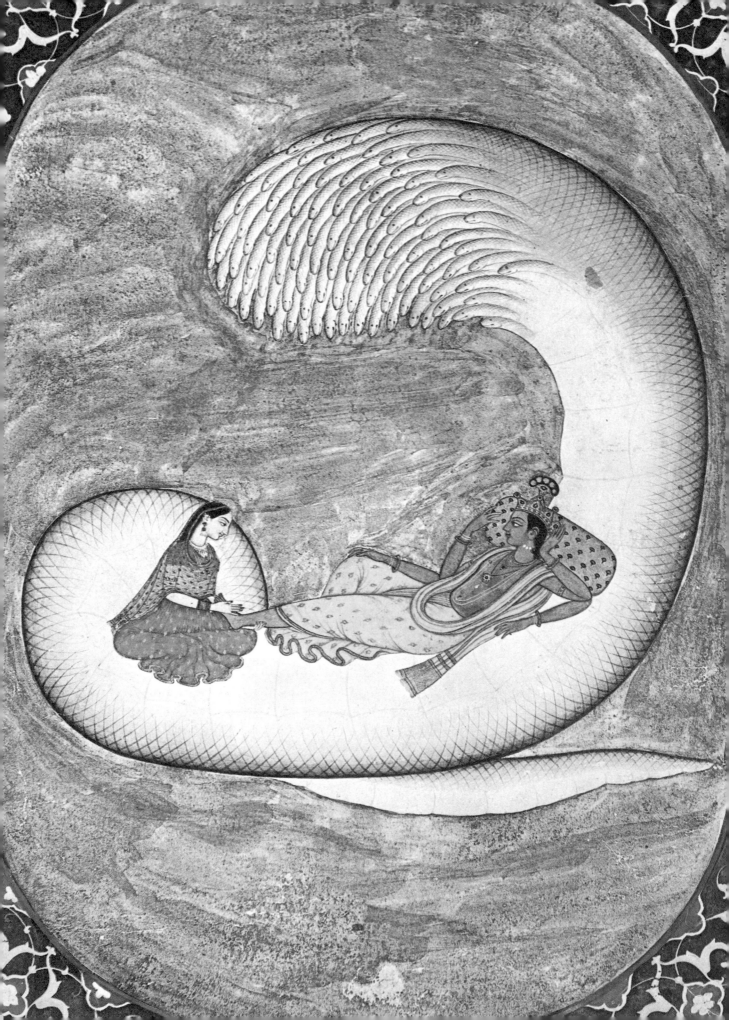

But did it? First, it is possible that in his boredom and lack of stimulus he spent far more time sleeping than he allowed for; secondly, when arriving at his final estimate he may have over-corrected his inner time, making too great an allowance for its exaggerations—more a gesture, at once manly and modest, than a mistake. And his Time-confusion was no worse than ours, and we are in the daylight, surrounded by clocks, calendars, and almanacs, to say nothing of mathematicians, physicists, and astronomers.

This brings me back to the point I made earlier, when I said that our inner sense of duration, psychological time, on the level on which it is usually examined, does not in my opinion take us very far into the problem of Time as I see it. And I realize that this statement needs some enlargement and clarification. Bergson, an original and bold if rather reckless philosopher, based a whole philosophy on the idea that outer or chronological time is unreal and that reality can be found only in our inner sense of duration (*durée*) or psychological time, the unceasing and creative flow of which we have an immediate apprehension. This real time, he insists, has nothing to do with space, and we must reject the almost constant temptation to "spatialize" our thought about Time. I do not propose to offer a technical criticism of Bergson (indeed, I am ill-equipped for any such attempt) but there are some remarks I must make, if only to forward our inquiry here.

In the first place, although some bad arguments may be based on the confusion of Time and Space, these two difficult concepts (which have so much in common) cannot be definitely separated, and we may arrive closer to the truth if we do not pull them entirely apart. We may make more progress if we take the risk of being accused of "spatializing" Time. It is better to arrive somewhere, even with shouts of disapproval ringing in our ears, than to be philosophically invulnerable but arrive nowhere. This is certainly true if our primary object is not to produce a faultless disquisition but to help other men to find a good life.

Secondly, in actual practice we do not regard our inner sense of duration, psychological time, as being real and outer chronological time as being unreal. When we may appear to do so we are knowingly indulging in exaggerated talk and throwing metaphors about. When, for example, I meet two old friends after a long interval and spend several delightful hours with them, and then declare I do not know where the evening has gone to, I have not really lost all sense and appreciation of clock time, have not proved it to be unreal; I know very well I have just passed an evening of it.

And when I have found myself trapped in a lecture on The Lancashire Coal Trade in the later 18th Century or on The Sociological Significance of the Small Group, and I say afterward that I was there for years and years, I am only talking for effect and am quite aware of the fact that I spent exactly an hour on that chair too small and hard for me. I am still accepting clock time as a reality and am merely playing with and coloring my account of it. And this is a very different matter from feeling conscious of *another kind of time*.

So here is my third objection to this enormous emphasis upon psychological time. It tends not to further but to sidetrack any real attempt to solve the problem of Time. It crams into one holdall, labeled *psychological time*, too many quite different sorts of experience, some of which may have nothing to do with chronological time. But this independence is rarely noted and granted. Clock time rules reality for most experimental psychologists, who are generally more anxious to be "scientific" than the professors of the exact sciences.

And what may easily get lost in this holdall of psychological time is the kind of experience (which undoubtedly exists) that seems to be clean outside chronological time and cannot be squeezed into the accepted temporal pattern.

To the Hindus, Time is divided into enormously long cycles of creation that correspond to the 10 incarnations of the god Vishnu (who appears in the story on p. 70). Left, an Indian drawing depicts Vishnu with his wife Lakshmi resting on the thousand-headed snake Sesha, in an interval between successive creations.

69

We are closer to what I have in mind when we remember all the myths, legends, and fairy tales in which there are more times than one, a relativity of times, and which have had for many of us, from our childhood onward, a strange fascination, as if behind the adventures and marvels a profound secret was being revealed.

I am thinking now of those tales of poets who were lured into some fairy kingdom and then returned to their own time and place, only to discover they were very old men whom nobody remembered; or of that medieval abbot who, on a spring morning in the fields, fell into an ecstatic trance and, when he walked back to the abbey, could not find there anybody who had ever heard of him, several centuries having passed. One of the best of these Time myths, dramatically illustrating relative times and degrees of illusion and reality, I owe to Professor Mircea Eliade, from whose *Images and Symbols* I now quote. The setting of the tale is of course Indian:

A famous ascetic named Nārada having obtained the grace of Vishnu by his numberless austerities, the god appears to him and promises to do for him anything he may wish. "Show me the magical powers of thy *māyā*," Narada requests of him. Vishnu consents, and gives the sign to follow him. Presently, they find themselves upon a desert road in hot sunshine, and Vishnu, feeling thirsty, asks Nārada to go on a few hundred yards farther, where there is a little village, and fetch him some water. Nārada hastens forward and knocks at the door of the first house he comes to. A very beautiful girl opens the door; the ascetic gazes upon her at length and forgets why he has come. He enters the house, and the parents of the girl receive him with the respect due to a saint. Time passes, Nārada marries the girl, and learns to know the joys of marriage and the hardships of a peasant life.

Twelve years go by: Nārada now has three children, and, after his father-in-law's death, becomes the owner of the farm. But in the course of the twelfth year, torrential rains inundate the region. Supporting his wife with one hand, holding two of his children with the other and carrying the smallest on his shoulder, Nārada struggles through the waters. But the burden is too great for him: he slips, the little one falls into the water; Nārada lets go of the other two children to recover him, but too late; the torrent has carried him far away. Whilst he is looking for the little one, the waters engulf the two others, and,

shortly afterwards, his wife. Nārada himself falls, and the flood bears him away unconscious. . . .

When, stranded upon a rock, he comes to himself and remembers his misfortunes, he bursts into tears. But suddenly he hears a familiar voice: "My child! Where is the water you were going to bring me? I have been waiting for you more than half an hour!" Nārada turns his head and looks: Instead of the all-destroying flood, he sees the desert landscape, dazzling in the sunlight. And the god asks him: "Now do you understand the secret of my *māyā*?"

Obviously Nārada cannot claim to understand it entirely; but he has learned one essential thing: he knows now that Vishnu's cosmic *māyā* is manifested through time.

3.

To deal justly with the metaphysics of Time we should need a book at least as big as this, and it would have to be written by somebody else. (As it is, I am already beginning to feel that even this book ought to be written by somebody else.) Some of the metaphysicians, from Leibniz and Kant onward, remind me of what happened to me once in a small shop in a French village. I had chosen what I wanted to buy, had paid for it, and was about to go when I decided that I had put down my spectacles somewhere in the shop. The woman there searched as earnestly as I did, and it was several minutes before we discovered, at the very same moment, that I was in fact wearing them, having absentmindedly put them on to look for them.

And this situation is not unlike one metaphysical idea of Time: We do not discover Time but bring it with us; it is one of our contributions to the scene; our minds work that way.

Another metaphysical solution of the Time problem is shortly described by Dr. M. F. Cleugh in the Introduction to her large and extremely conscientious study *Time: and its importance in modern thought*. After indicating the appalling complications of the subject, Dr. Cleugh goes on:

Confronted by all these difficulties, what is to be done? One very simple and obvious way out, and one which has been popular with idealists of all ages—Parmenides, Plato, Spinoza, Hegel, Bradley and McTaggart—is to say that time is riddled through and through with contradictions, and hence cannot be real. The major premise of the argument is that the Real (note the capital) cannot be self-contradictory, or possess contradictory characters. It is a summary solution of the difficulty and one for which there is much justification. Nobody who has contemplated the awe-inspiring accumulation of problems given above can be out of sympathy with it. But *merely* to say that because time is self-contradictory it must be appearance only, is, so far from solving the problems, not even an answer to them. It is doubtless interesting to know that time is not real, but only apparent, but it is not sufficient. The denial of reality to time avoids the main problem, for you have still to explain the *appearance* of time. McTaggart is almost alone among idealists in realising this.

Let us see then what we can make of McTaggart, who may be out of fashion now in philosophical circles but whose assault upon this Time bastion, now more than half a century old, has not been forgotten, and indeed is recognized to be the most formidable attempt yet made from the idealist's position. But first let me admit that I have another and more personal reason for preferring McTaggart. He was still lecturing when I was up in Cambridge, and though philosophy (known there as "moral sciences") was not one of my subjects, I used to attend his more elementary lectures on metaphysics. While listening to him I could never find a flaw in his lucid and highly ingenious arguments, but rarely believed anything he had told me once I was out of his presence. (I am no logical positivist, as this book will make plain, but I always feel that metaphysicians like McTaggart are playing a wonderful game that has nothing to do with truth and reality.)

His presence was delightful: He had a curious high voice, a large moon-baby face with spectacles on the end of its nose, and he held his head to one side and stared up at the ceiling while he illustrated his argument by references to pink elephants. He was one of the great originals of Cambridge. Some odd disability gave him a crab-like walk, and one met him coming sideways around buildings, like a sheriff about to shoot it out with the bad man in a Western. He had a passion for reading novels, any kind,

thousands and thousands of them; and it was said he kept a special edition of Disraeli's novels to read in his bath. He believed in human immortality, was a staunch supporter of the Church of England, and, having cheerfully argued God out of the universe, he was an atheist. Now let us see how this extraordinary man dealt with Time.

He divides what we discover in Time into two series, A and B. The A series is that of past-present-future. The B series is responsible for the relation "earlier than" or "later than," and any distinctions made in this series must be permanent, because if an event is *ever* earlier than another event it is *always* earlier. Nevertheless, the B series is derived from the A series: There cannot be time without change, and change cannot be found in the B series (the "earlier than" and "later than" being permanent and not involving change). It is, then, the A series of past-present-future that is essential to Time.

But now we meet a number of contradictions. To begin with, past, present, and future cannot be regarded as "qualities" of objects or events. A thing cannot be past or present in the same sense in which it can be green or soft or heavy. Past, present, and future then are *relations* in which things stand to some particular "term," which is essential to giving these things their temporal characteristics. But this "term" cannot be in Time, where past, present, and future are already; to distinguish between them it must be outside Time. And no such "term" can be found.

Furthermore, past, present, and future are clearly not compatible characteristics, yet every event has to have all three. And we cannot escape this contradiction, as we might in dealing with qualities, by saying that an event may have these incompatible characteristics "at different times" because that either takes us around in a circle or plunges us into an infinite regress. Therefore the A series, on which the B series depends, is self-contradictory, and Time is unreal. It is an appearance, and as such is real, but things appear to us to be in Time because we misperceive them.

This brings McTaggart to a third or C series, which is a genuine series and not self-contradictory (his account of it is too elaborate to be quoted here) and which leads him to declare: "My view is, then, that whenever a self (or a part of a self determined by determining correspondence) appears as being in time, it is divided in another dimension besides those of its determining correspondence parts, and the terms in this fresh dimension form the C series." His conclusion is that Time is unreal, but has some reality for us as an appearance owing to an "erroneous perception" from which in this life we cannot escape.

I have not offered the foregoing as an adequate account of McTaggart's metaphysics of Time. (The curious reader should try his *The Nature of Existence* or, failing that, Dr. Cleugh's examination of him in her book previously mentioned.) My chief reason for even trying to make the attempt was to offer an illustration of the appalling complexity and difficulty of any philosophical or metaphysical approach to the problem of Time. It is as if a riddle, when partly solved, only revealed other and still more teasing

riddles; as if any departure from the well-trodden road of common sense hurried us at once into mazes and bogs and dead ends.

Yet if we are both curious and deeply concerned about our life and our destiny, that road, so safe and easy at first, will not take us anywhere worth arriving at, not even within sight of any rewarding goal. So we shall have to leave it and dodge about and scramble on as best we can. And why this should be so, why Time should be so elusive and enigmatic, I must now try to explain.

The past, present, and future of the Elizabethan courtier Sir Henry Unton (center) illustrate McTaggart's Time series A in the form of one man's life from birth (right) to death (left). In terms of this theory, past, present, and future are not "qualities" of Sir Henry, but relations that can be distinguished only by reference to some point outside Time—which therefore becomes meaningless. Also, every event must have a past, present, and future; yet these factors cannot exist simultaneously. So, McTaggart concluded, Time is only an appearance, not a reality.

"The wildest theory of fantastic dimensions seemed to me to have a little more sense in it than the conventional view of Time, which nearly everybody I knew accepted without question." I wrote that nearly 30 years ago. Possibly the Time situation has eased a little since then. The wider acceptance of the theory of relativity and its space-time continuum, speculations about space travel, science fiction with its tricks with Time, may have disturbed and blurred that conventional view.

We are able to believe that an observer on a distant planet, using his two heads and five tentacles for advanced astronomy, might possibly be taking a look at the Battle of Hastings and wondering which side will win. And we know that those of our astronomers who are, so to speak, straining their eyes and ears the hardest are really considering the unimaginably distant past of the universe; they are staring through space at Time. In fact, we do it ourselves whenever we glance at the night sky, where the twinkle is in our present time but the twinkler may be as dead as Queen Anne.

Here is a note, in passing, on scale. Most of the scientific texts I have read, or tried to read, during the last 10 years, have dealt with matters outside the medium scale in which our thought is at home. They have been concerned with things either incredibly vast or inconceivably tiny, with the movement of galaxies or the whirl of electrons and neutrons. Now the table on which I am writing fits comfortably into our scale and time. But as soon as we move outside our scale, this table vanishes. From the galactic point of view it simply does not exist. On the other hand, to its electrons and neutrons this table may appear to be an unimaginably vast, ever-enduring, multi-dimensional universe.

Now if we demand that any idea of Time should include all these things so far outside our scale, from Andromeda to a proton, I believe we shall begin in confusion (that is inevitable) and soon end in chaos and madness. I am thinking then of Time within our scale, somewhere between galaxies and electrons, the scale in which, somewhere near the middle, this table remains a table and I can continue writing on it.

In spite, however, of that slight easing of the situation, and all the pieces on space travel and all the paperback science fiction, the conventional view of Time has hardly shifted an inch since I wrote about it, nearly 30 years ago.

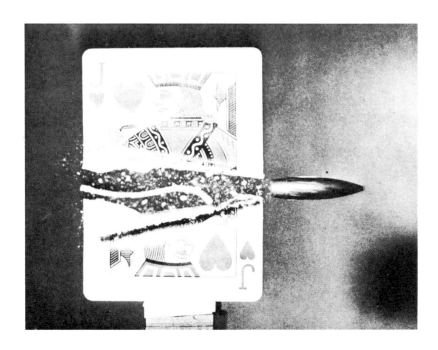

The conventional view of Time is that it is made up of a succession of "Nows," or instants. Left, an expression of a "Now": A high-speed photograph of a bullet slicing through a playing card.

Above, a mass of lines and shapes, among them a whale and (right) a lizard, that were drawn in the Peruvian desert by a pre-Inca civilization—perhaps as part of a giant calendar. These animal figures are so vast that the people who drew them must have been able to grasp only a fraction of the drawing at a time: They must have progressed in much the same way as, for many people, the "Now" (or present) seems to move steadily forward—like Priestley's image of a slit in a moving barrier (p. 76). Yet just as these unknown artists could visualize the whole drawing even without aerial photography, so we can perceive (and experience) more than just the present instant: Our awareness includes the past *and* the future.

That is why I declared that confusion is inevitable. Common sense still tells us that everything is real only when it is Now, in the present moment. Whatever is not-Now does not exist. Reality is served to us in these thin slices of Nows. It is as if we watched a film of which each successive frame magically arrived from nowhere to turn into nothing. And common sense is so wedded to this idea of our life as a tightrope balancing act, between nothing and nothing, nowhere and nowhere, that if a man claims to have caught a glimpse of the future and a woman says she has seen the past, common sense will denounce them as self-deceivers or charlatans. It has settled down with the notion that everything is real only when it is Now, so let the world keep on destroying and re-creating itself every fraction of a second or so: It is doing a grand job.

There is an opposite view. Those who take it have firmly turned their backs on all that destruction and recreation, that endless rushing from nowhere to nothing. In this theory, which has gone to the other extreme, everything is solidly there, whether we call it past, present, or future. We experience things in time because our Now, so to speak, goes steadily forward, as if we were traveling through a dark landscape with a searchlight, or we were staring at a bright scene through a slit in a moving barrier. We invent Time to explain change and succession. We try to account for it out there in the world we are observing, but soon run into trouble because it is not out there at all. It comes with the traveling searchlight, the moving slit.

And now there seems no reason why that man should not catch a glimpse of the future or that woman see the past: They are there to be seen. There is, however, one snag. How can the searchlight, the moving slit, steadily revealing what we call the present, be manipulated to light up, to offer a slitful, of the future so much further on or of the past that has been left behind? If Time is really our name for this steady forward movement of consciousness over the fixed scene, then what are we going to call this jumping about?

I have noticed that, whenever I am offered this block universe, all solidly there to be discovered in the light of our successive moments,

I am never called anything but an Observer. I am like a man on a moving roadway in some World's Fair. I am a rustic in the city for the day, staring at the traffic, the lights, the skyscrapers. But though I know this role, I rarely play it. I may be no maker of history but I have not spent my life simply observing. I am not just looking at the scene, *I am in it*, doing this, doing that.

And if I am going to be told that my idea that I make choices, take action, interfere, possibly change the future, is all an illusion, then I shall want to know how this block universe, this frozen history, came into existence, who shaped it, who colored it, and what is the point of this vast idiotic conjuring trick. A consciousness that is no more than a policeman's lantern moving along a back alley—and indeed much less, because no action can follow from it—is not worth having.

No doubt if we make an imaginative effort we can go along with the mathematicians and physicists and see our lives as world lines in a space-time continuum. "Think of yourself," one scientist writes, "as a four-dimensional figure, a kind of long rubber bar extending in time from the moment of your birth to the end of your natural life." (Not many years ago, if a scientist had been asked to explain the ancient Indian term "the long body of man," he would have replied that old Oriental nonsense was well outside his field.) But if we are trying to explain Time, we shall have to be very careful about these world lines, these long rubber bars extending in time. For example, if we imagine our consciousness traveling along these world lines like a spark in a fuse, and if Time is now the length along which it travels, then we must be careful not to begin timing this movement because that would involve bringing in another time, then another to time that, and then another, and off we go into an infinite regress.

On the other hand, if our world lines are not already there, so to say, but grow with our lives—our consciousness moving on like the luminous head of a snake—then this growth, this movement, would be in time as a fourth dimension; but when we try to imagine this progress we seem compelled to put ourselves outside this time, taking a viewpoint that demands another

The hands of an ordinary clock moving through a succession of minutes provides the most common image of time's passing. But not all clocks behave conventionally: For example, the hands of the 17th-century Hebrew clock beneath the town-hall clock in Prague (left) move backward, while those of the 18th-century French clock (below) describe a semicircle; as soon as they reach the right-hand figure six they go back and start again from the left. In the same way, not all experience of Time can be explained in terms of the common-sense idea of passing time.

dimension, another time. This does not worry me; I am a crank anyhow. But it ought to begin worrying those scientists who offer us these world lines in a space-time continuum, these lives as long rubber bars extending in time, like tossing a biscuit and a rubber bone to a dog, and then tell us not to talk nonsense about a Time problem. For it is still here, staring them in the face, a crouching Sphinx.

The appalling difficulty of examining the Time problem seems to me to be chiefly due to the fact that Time either changes into something else or quietly disappears from the examination room. Trying to keep it fixedly in view is like playing Wonderland croquet. Dr. Cleugh, a severe logician, frequently complains in her book (and she is quite within her rights to do so) about other people's *hypostatization* and *spatialization* of Time. Now I am ready to plead guilty to both these metaphysical crimes, but I am also ready with a case for the defense.

To hypostatize Time—and the prosecution will call that capital T as a witness—is to give it actual existence and to cheat in a few attributes

When we hypostatize Time, we give actual existence to a concept abstracted from our experience of succession. The process can be represented by the visual analogy of a "frozen moment." Such an image is provided by the lava-encased body of a Pompeiian (below): The moment that he fell (when Vesuvius erupted in A.D. 79) is preserved permanently—thus is abstracted from reality.

Before the invention of the camera, artists often wrongly depicted "frozen time," as in the drawing above left : A galloping horse never achieves such a position. But the painting by a 17th-century Italian artist of an explosion in a church (below) has—by a remarkable imaginative effort on the part of the artist—depicted an instant of time as accurately as a high-speed camera.

Another example of a "frozen moment" as an unreal abstraction. The photographs (above) can be said to represent an observation of a galloping horse. Any *one* of the four poses might be selected to represent the concept "galloping horse." Yet such an arrested pose has no *separate* existence in reality : It has merely been abstracted from our experience of succession.

to which, as a concept, it has no claim. We abstract a concept from our experience of succession, call it Time, then turn it into an immeasurably vast container, into which everything goes; and then we wonder why we cannot make head or tail of it. We are worse than those ancient peoples who believed that behind all change and succession was a special god, more real than they were. At least they ended with a god, who could be worshiped and might be placated, whereas we are trying to measure and analyze a ghost-container.

But did I not show people successfully measuring Time in my first chapter? The answer is that I did not, for it was time, not Time, they were measuring. The big one, with the capital T,

refuses to be confined within the clocks and calendars. Just when it appears to be pinned down, ready for tests and analysis, it escapes from the experiment and joins the experimenter. But is not this a kind of hypostatization? It is, indeed, but I said I was ready to plead guilty.

As for the other metaphysical crime, "spatialization": Standing as I do, well outside metaphysics, I am determined to be tougher in my defense and even declare that I should not be charged. But I am reserving this defense for the final section of the next chapter.

Now if, as we go along, I appear suddenly to feel some uncertainty, to weaken in my grasp of our subject, it will be because I am remembering about hypostatization, reminding myself that

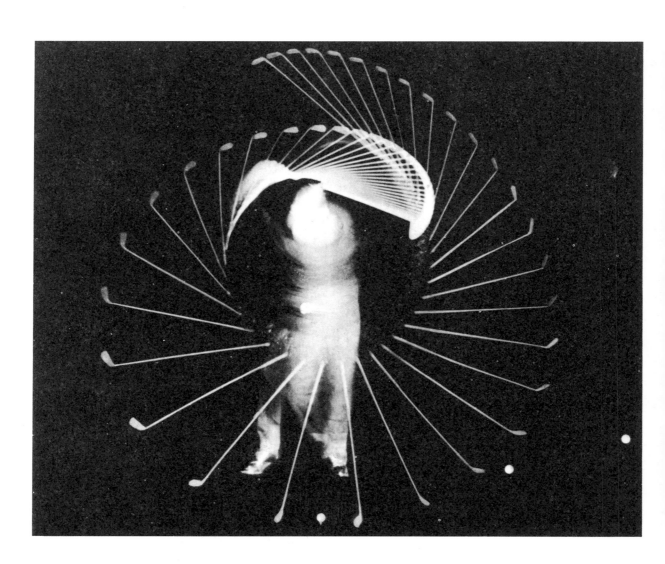

Time is not a thing in actual existence with qualities and attributes. If I seem to suggest that it is, I am doing this for convenience of reference. I realize that Time is a concept abstracted from our unavoidable acquaintance with succession and all our *befores* and *afters*. I regard it as a condition of our experience; and if I refer, as I certainly will, to different times, I am referring to different temporal conditions for different kinds of experience.

But why is it so difficult to capture Time? Well, here is an analogy that does not seem to be entirely fanciful. We see two men bending over a chess board or 22 men hurrying about a football field. Now we cannot really understand what these men are doing unless we know something that is not to be found on the board or in the field. What are invisible there and yet give the board or the field meaning are the rules of the game. On our various levels of existence, we may play different games under different sets of rules; and so we may discover different kinds of time offering us different kinds of experience.

But these times, which could have other names, fall away from Time itself, which remains in our eyes as we look, in our minds as they record, refusing to be an object that can be detached and captured. In this respect, pursuing Time, we are like a knight on a quest, condemned to wander through innumerable forests, bewildered and baffled, because the magic beast he is looking for is the horse he is riding.

Left, a multiple-exposure photograph records a golfer's swing. Yet the individual positions during the swing exist only for an instant, so to give each of them permanent form in this way is really a kind of falsification. But as such a photograph may help a golfer to analyze his swing, so the hypostatization of Time is a convenience to anyone studying or writing about Time.

Different temporal conditions occur for different kinds of experience. As later chapters will show, one of the most striking differences appears between the temporal conditions of our waking experience and those of our sleeping (or dreaming) experience.

3   Time among the Scientists

The laboratory at Zurich where, during the early 1900s, Albert Einstein first formulated his revolutionary theories of relativity. Einstein's theories marked the final liberation of scientific thought from the rigid framework imposed by Newtonian physics and opened the way toward a new examination of space and Time.

the question is meaningless, belonging to the angels-on-the-needle category. I do not say that first-class scientists hold this opinion, which could bring science itself to a halt; but it is not uncommon in positivist circles, where theories about Time are regarded with mixed suspicion and contempt.

Among scientists it is the physicist who is most closely concerned with Time, and among physicists it was one of the earliest and greatest of them, Newton, who produced (not without a certain flavor of divinity) a definition of Time that dominated scientific thought for many generations. Here, a quotation from G. J. Whitrow's excellent *Natural Philosophy of Time* is appropriate:

"Absolute, true and mathematical time," wrote Newton, "of itself, and from its own nature, flows equably without relation to anything external." This famous definition which appears at the beginning of the *Principia* has been one of the most criticised, and justly so, of all Newton's statements. It reifies time and ascribes to it the function of flowing. If time were something that flowed then it would itself consist of a series of events in time and this would be meaningless. Moreover, it is equally difficult to accept the statement that time flows "equably" or uniformly, for this would seem to imply that there is

something which controls the rate of the flow of time so that it always goes at the same speed. But if time can be considered in isolation "without relation to anything external," what meaning can be attached to saying that its rate of flow is *not* uniform? If no meaning can be attached even to the possibility of non-uniform flow, then what significance can be attached to specifically stipulating that the flow is "equable"?

There is, as I have already suggested, a flavor of divinity about this definition of Newton's; there falls on it, we might say, the shadow of an earlier kind of thinking. This absolute time of his is quite independent of the sequence of events, though it is, of course, from the sequence of events that we in fact derive our idea of Time. We have here a suggestion of a divine order to which nature as we know it is only an approximation; but unless we are ready to claim direct revelation, we can know nothing about a divine order except through nature. And one of the dominating ideas of the 18th century—the idea of a deity who set the machinery of the universe in motion and then retired—was in great part derived from Newton.

The Newtonian reign in physics lasted for at least 200 years; as Professor G. Gamow has said:

So strong was the belief in the absolute correctness of these classical ideas about space and time that they have often been held by philosophers as given *a priori*, and no scientist even thought about the possibility of doubting them. However, just at the start of the present century it became clear that a number of results, obtained by most refined methods of experimental physics, led to clear contradictions if interpreted in the classical frame of space and time. This fact brought to one of the greatest contemporary physicists, Albert Einstein, the revolutionary idea that there are hardly any reasons, except for those of tradition, for considering the classical notions concerning space and time as absolutely true, and that they could and should be changed to fit our new and more refined experience. In fact, since the classical notions of space and time were formulated on the basis of human experience in ordinary life, we need not be surprised that the refined methods of observation today, based on highly developed experimental technique, indicate that these old notions are too rough and inexact, and could have been used in ordinary life and in the earlier stages of development of physics only because their deviations from the correct notions were sufficiently small. Nor need we be surprised that the broadening of the field of exploration of modern science should bring us to regions where these deviations became so very large that the classical notions could not be used at all.

We shall come in the next section to what was discovered in that broader field of exploration. Here I should like to make one or two points before moving on. The developments noted by Professor Gamow above were, I believe, responsible for an important change in the relation between science and the general public. Thus, those "classical notions," which had to be discarded, were not difficult to understand, with the result that, throughout the 18th century and most of the 19th, there was no great gap between the physicist and the ordinary educated layman. Indeed, these were the years when the amateur scientist flourished. But with the coming of more "refined methods of observation," more "highly

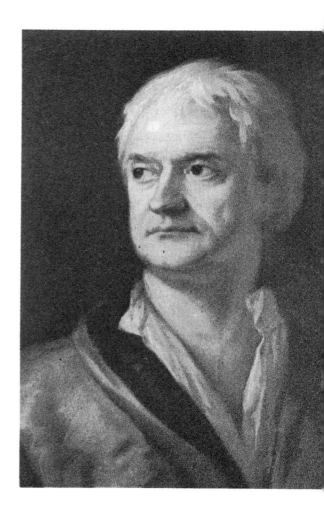

Above, a portrait of Sir Isaac Newton (1642–1727), the great physicist whose definition of Time remained more or less unchallenged for 200 years. Left, a line represents Newton's idea of Time as an "equable flow," whose uniformity is suggested by six stopwatches above a sundial. Variations of this statement put forward by Newtonian physics still underlie the conventional view of Time as an unvarying flow that can be measured by clocks and sundials—despite, to many people, the failure of this narrow view to explain all their different experiences of Time.

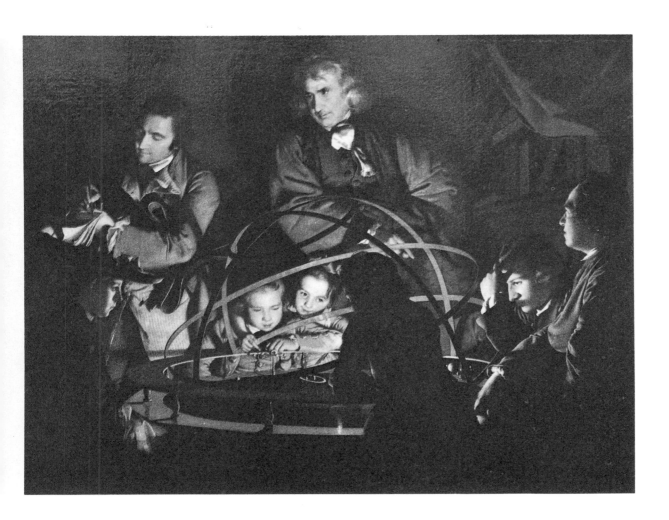

Left, a 16th-century woodcut shows the Greek mathematician Pythagoras (right) and the Roman philosopher Boethius inspired by the muse of arithmetic. In ancient times, most scientific thought included a mystical element—an element that finally disappeared in the "Age of Enlightenment," when the scientific laws postulated by Newtonian physics transformed nature's mystery into nature's "machinery." Above, in *The Orrery* (by the 18th-century British artist Joseph Wright) a party examines a clockwork machine that demonstrates planetary movements.

developed experimental technique," an increasing specialization, together with the breakdown of "the classical frame of space and time," the scientist and the layman began to lose sight of each other.

It is true that ours is the age of science, that far more science is taught today than ever before, that popular expositions of science are on the increase; but it is also true (especially in physics and its companion studies) that, while science speaks with more authority than ever, it speaks from mysterious heights and depths beyond the sight and understanding of the lay public. It seems to appeal no longer to common sense, and now appears remote from "our daily experience of the world" in which my clever young friend declared it is "firmly rooted." (For example, Newton's idea of absolute time flowing equably was soon part of common thought, whereas after half a century, and after countless books, papers, lectures, and so on all devoted to it, Einstein's theory of relativity has never taken hold of the public mind.)

The result has been that to most people science seems a mystery, remote and awe-inspiring, out of which certain dogmas appear, like tablets of stone brought down from the shrouded mountain. And there are further consequences we shall discover in due course.

2.

The physicists in fact effected two breakthroughs: the one that gave us relativity, and the other that took them into the unimaginably microscopic world of the atom and the subatomic particles. Both in their different ways, the one on an immensely large scale, the other on a fantastically small one, have some concern with Time; I will try to limit my account of them to this term of reference, so far as it is possible.

Now I am old enough to remember, when I was being taught some elementary science at school, hearing much talk about "the ether." What my instructors did not know was that Einstein was already busy blowing this mysterious stuff away for ever. Nor did they seem to know (though it had happened 20-odd years before) the experimental work of the American physicists Michelson and Morley, which had proved that the earth's motion had no influence whatsoever on the velocity of light. This remained constant whether going with or against the earth's motion. And if classical physics were right, then this behavior of light was all wrong. But further experiments only proved that it continued to behave badly. And then Einstein arrived on the scene.

As part of his Special Theory, Einstein stated that the velocity of light is always constant relative to the observer, whatever his motion might be. Time is not absolute but relative to the position of the observer. This idea, in the Special

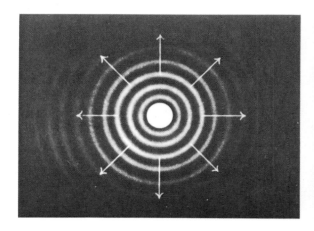

Below, a three-dimensional picture of light's motion: Apparently light is propagated in the form of successive "areas" of electromagnetic energy moving in a series.

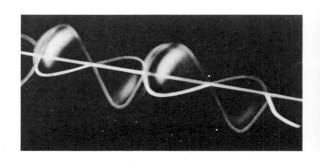

The American physicists Edward Morley (right) and Albert Michelson (below), whose experiments with the velocity of light first led scientists to question the generally accepted idea of the "luminiferous ether" —the substance that, it was thought, permeated the universe and through which light traveled. Below left, an illustration from a popular science book of 1900 shows light radiating outward through the ether like ripples in water when a stone is dropped.

The physicists Michelson and Morley set out to prove that the speed of light is totally unaffected by the earth's motion. Here three pictures show (top) light traveling in the same direction as the earth, and (top right) light traveling against the earth's motion. In both cases, it was proved that light traveled at a constant rate (right). Below, a drawing of the rotating table mounted with mirrors that was set up by Michelson and Morley for their experiment.

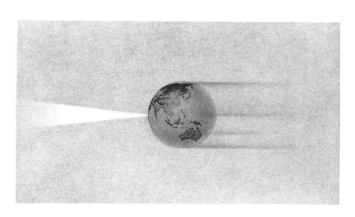

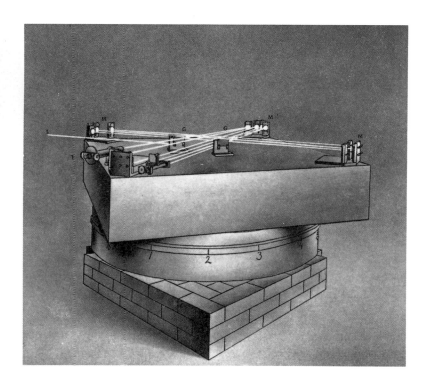

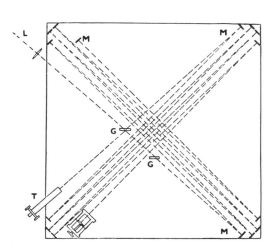

Above, a diagram shows how light from L was partly reflected and partly passed by mirrors (at M and G) in criss-cross beams. Readings taken at T indicated that there was no appreciable change in the speed of the light beams whatever the observer's position.

Theory, was expressed in two forms: first, in the time-dilatation effect; secondly, in its stress on the fact that two observers at different distances from an event will see it at different times.

This second form of the basic idea is the easier to understand, for if a ray of light takes time to reach our eyes, it will take longer to reach them the farther we are away from it; so two people a large distance apart could justifiably disagree about the date of an event. This is clearly illustrated by a figure and an explanation based on it by James A. Coleman in his *Relativity for the Layman*. This figure

shows the earth, the star Betelgeuse in the constellation Orion, the Hunter, and Aldebaran in Taurus, the Bull. Betelgeuse and Aldebaran are 300 and 53 light years respectively, from the earth. Also, Aldebaran is about 250 light years from Betelgeuse. Now suppose there is a "blow-out" in Orion on the night of March 17, 2000, caused by Betelgeuse exploding. This date, and all the dates mentioned here, refer to our method of keeping track of time on earth. We on earth would not see the blow-out on that date, since Betelgeuse is 300 light years away, which means it would take 300 years for the light waves of the explosion to reach us. This is the only way we would learn of the explosion. The date for us would be March 17, 2300. Somebody on Aldebaran, on the other hand, would see the explosion on March 17, 2250, since Aldebaran is 250 light years from Betelgeuse. . . .

This is clear enough, though we must remember that March 17, 2000, the date of the explosion, is an *earth date*, which would only be arrived at by inference on or after March 17, 2300; and that in terms of the universe there is no date for the explosion, any idea of the simultaneous being impossible. So the idea of Absolute Time can no longer be reasonably entertained.

It has been pointed out that, if we can imagine observers moving away from the earth at speeds comparable to that of light, some curious time effects would appear. If they moved at exactly the speed of light, let us say, on Christmas Day, any earth light signals would remain the same so that they would, so to speak, stay in Christmas Day. But if they moved faster than the speed of light (a most unlikely event), the time would appear to be reversed for them,

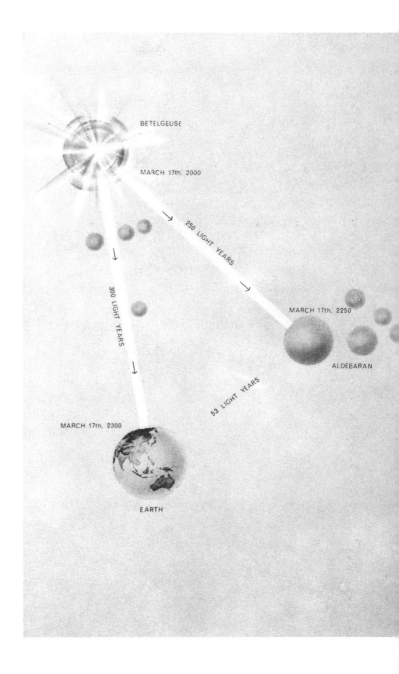

Einstein's Special Theory showed that Time (unlike the speed of light) is relative to the position of the observer. For example, should an explosion take place on Betelgeuse on March 17, 2000 (earth time), it would be "seen" 250 years later on Aldebaran and 300 years later on the earth.

because they would overtake the signals sent out by Christmas Eve and then by the day before Christmas Eve, so that earth time would appear to be running backward. Speeds faster than that of light can be imagined, but are not likely to be achieved, except in thought. But if thought moves, what is its speed?

However, we must return to Einstein's Special Theory, and the time-dilatation effect. This is more complicated than the other form of the idea. It asserts that two observers, moving at a large velocity (near that of light) relative to one another, will each see the other in slow motion, as if everything to do with him were running "behind time." Moreover, in his General Theory (which came out 11 years later than the Special Theory), Einstein also showed that mass affected the rate of time, so that if, for example, the earth had been larger, time would have been slower on it.

As Mr. Coleman points out in the book already mentioned, a clock running at a certain rate on earth would run slower on Jupiter and slower still on the sun. And it was discovered that a second of the sun's time would correspond to $1 \cdot 000002$ earth seconds. After just under six days the sun clock would be one second behind the earth clock, if we suppose them to have been synchronized at the start of the first day. No such clock experiments were possible, of course, but Einstein's first predictions were proved to be correct by complicated comparisons between the vibration-times of atoms on the sun with those of similar atoms on the earth.

So Einstein showed, first in his Special Theory, that the time of an event was not absolute but dependent on its position in space, and, later in his General Theory, that time was also relative to mass. He took us from a three-dimensional world existing in Absolute Time to a four-dimensional world and a space-time continuum. As Professor Dirac observes in *The Evolution of the Physicist's Picture of Nature*:

We have, then, the development from the three-dimensional picture of the world to the four-dimensional. The reader will probably not be happy with this situation, because the world still appears three-dimensional to his consciousness. How can one bring this appearance into the four-dimensional picture that Einstein requires the physicist to have?

What appears to our consciousness is really a three-dimensional section of the four-dimensional picture. We must take a three-dimensional section to give us what appears to our consciousness at one time; at a later time we shall have a different three-dimensional section. The task of the physicist consists largely of relating events in one of these sections to events in another section relating to a later time. Thus the picture with four-dimensional symmetry does not give us the whole situation. This becomes particularly important when one takes into account the developments that have been brought about by quantum theory. Quantum theory has taught us that we have to take the process of observation into account, and observations usually require us to bring in the three-dimensional sections of the four-dimensional picture of the universe. . . .

We have now arrived at what I previously described as the second breakthrough—into the unimaginably microscopic world of the atom and the subatomic particles, the protons, electrons, positrons, neutrons, and whatever else is turning up in that strange world. There, everything is utterly bewildering to the layman, and by no means clear and determined for the nuclear physicist. So Professor Norwerd Russell Hanson, in his *Concept of the Positron*, after discussing the diameter of an electron, can write:

If theory just tolerates the electron having a diameter, it ought also to have a shape. What shape? Is it spherical? Punctiform? There is at present no experimental information enabling one to form any consistent geometrical model of the electron. . . .

What about the electron's solidity? Again, neither theory nor experiments help. Some theoreticians feel the concept to be meaningless; others remark that

The German-born mathematician Albert Einstein (1879–1955), whose theories of relativity revolutionized physics. In his first (or Special) Theory, Einstein demonstrated that Time and space are not absolute but relative to the observer; later he also showed (in his General Theory) that the rate of time is affected by mass.

the deeper electron penetration proceeds, the more difficult a decision becomes, because of the myriad new particles created by the probing particle and the target-electron. Still others think the particle may have a "solid" central core where some current theories break down. . . . If this is not the case, then the electron can only be described as a cloud of virtual particles plus a central bare point charge. . . .

A great deal is known about the particle, but what is known seems incompatible with classical ideas about matter. Thus, while one can always speak of the state of a classical particle, i.e. its simultaneous position and velocity, nothing like this can even be articulated in quantum theory. . . . This has implications: is it that physics is just not yet in a position to determine electron states? No. Quantum mechanics is the only theory through which electrons can now be understood at all; that theory excludes the very possibility of forming a consistent concept of an electronic particle's state. Either we speak precisely of its position, or of its momentum, but not precisely of both at once. Similarly we can speak of a person then as a bachelor, and now as married. But we cannot speak of him as being at once married and a bachelor; the reason is analogous to that within microphysics.

Some of us, entirely irresponsible in such realms as these, have often secretly longed for the breakdown of *either/or*, in which Western Man has placed such trust, and now in microphysics it appears to have happened.

Later, Professor Hanson observes:

Matter has been dematerialised, not just as a concept of the philosophically real, but now as an idea of modern physics. Matter can be analysed down to the level of fundamental particles; but at that depth the direction of the analysis changes, and this constitutes a major conceptual surprise in the history of science. The things which for Newton typified matter—e.g. an exactly determinable state, a point shape, absolute solidity—these are now the properties electrons do not, because theoretically they cannot, have.

In this mysterious realm, it appears, the experimenter cannot help but interfere with the experiment, and one can know the speed of a particle or its position but never both at once. So this is no place in which to discover any final meaning of space and time—or, indeed, cause and effect. We have moved a long way from the Time of classical physics long accepted by common sense.

On the other hand, it sometimes seems as if these startling ideas of the quantum physicists have become a new orthodoxy. There is, so to speak, a new kind of common sense, statistical and opposed to any search for a meaningful picture of the atom. However, we can range bigger guns on to this target, such as those of Prince Louis de Broglie, in his foreword to Professor Bohr's *Causality and Chance in Modern Physics*:

We can reasonably accept that the attitude adopted for nearly 30 years by theoretical quantum physicists is, at least *in appearance*, the exact counterpart of information which experiment has given us of the atomic world. At the level now reached by research in atomic microphysics it is certain that the methods of measurement do not allow us to determine simultaneously all the magnitudes which would be necessary to obtain a picture of all the classical types of corpuscles (this can be deduced from Heisenberg's uncertainty principle), and that the perturbations introduced by the measurement, which are impossible to eliminate, prevent us in general from predicting precisely the result which it will produce and allow only statistical predictions. The construction of purely probabilistic formulae that all theoreticians use today was thus completely justified.

However, the majority of them, often under the influence of preconceived ideas derived from positivist doctrine, have thought that they could go further and assert that the uncertain and incomplete character of the knowledge that experiment at its present stage gives us about what really happens in microphysics is the result of a real indeterminacy of the physical states and of their evolution. Such an extrapolation does not appear in any way to be justified.

It is possible that looking into the future to a deeper level of physical reality we will be able to interpret the laws of probability and quantum physics as being the statistical results of the development of completely determined values of variables which are at present hidden from us. It may be that the powerful means we are beginning to use to break up the structure of the nucleus and to make new particles appear will give us one day a direct knowledge which we do not now have of this deeper level. To try to stop all attempts to pass beyond the present viewpoint of quantum physics could be very dangerous for the progress of science and would . . . be contrary to the lessons we may learn from the history of science.

This teaches us, in effect, that the actual state of our knowledge is always provisional and that there must be, beyond what is actually known, immense new regions to discover. Besides, quantum physics has found itself for several years tackling problems which it has not been able to solve and seems to have arrived at a dead end. This situation suggests strongly that an effort to modify the framework of ideas in which quantum physics has voluntarily wrapped itself would be valuable.

And indeed I have no doubt it would. But here I propose to call a brief halt, if only to give ourselves a rest from the relentless prose of the scientists and their philosophers.

3.

When, in the chapter before this, I ventured a few remarks about scale, I was preparing the reader for what would emerge in this chapter, among the scientists. The idea of Time in classical physics (and the one long accepted by common sense) has been successfully challenged; the Newtonian reign has been brought to an end; but this has happened only on scales far removed from the world and the life of man. Thus, in the arguments for and examples of relativity, we are asked to consider light signals from galaxies, clocks millions of miles apart, velocities beyond our ordinary comprehension, with everything on so huge a scale that we ourselves shrink out of existence. My table and I and you are simply not there. And clearly, while some scientists have now given us plenty to worry about, the fact that a second of the sun's time would correspond to 1·000002 earth seconds is one of the least of our worries. I am not sneering at what was an important discovery. I have now almost a vested interest in anything that contradicts the idea of Absolute Time. And the Special

In order to formulate the effect of motion on the measurement of space and Time, Einstein decided to treat them together as aspects of a four-dimensional space-Time continuum. Of these four dimensions, we perceive only the three dimensions of space. But an idea of the fourth dimension of Time can be obtained from the diagram, right. The circular flight of an airplane (at the base of the drawing) is photographed from below at intervals of a minute. The photographic plates are developed and stacked: They then represent a three-dimensional continuum—two dimensions of space and one of Time. Next, a line is drawn connecting the airplane's successive positions— a line that, rising in a corkscrew through the continuum, will show the airplane's motion through both space and Time.

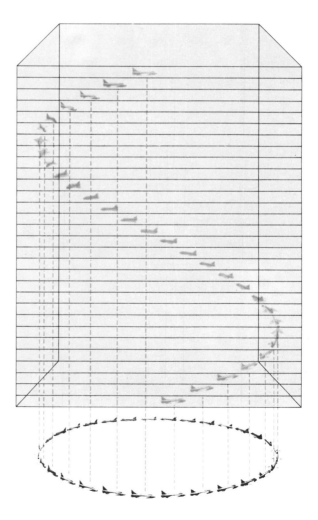

Above, Einstein reads a paper on his theory of relativity—a concept that was generally considered to be extremely difficult to understand. Below, a 1930 cartoon by Britain's George Strube (of Einstein talking to "the ordinary man") reveals a simplified way of looking at relativity. Einstein's work also aroused great controversy: Right, a montage of hostile comments by the press at the time of Einstein's arrival in America in 1921.

WELL, IT'S LIKE THIS,—SUPPOSING I WERE TO SIT NEXT TO A PRETTY GIRL FOR HALF AN HOUR IT WOULD SEEM LIKE HALF A MINUTE,—

BRAFFO! YOU THE IDEA HAF!

BUT IF I WERE TO SIT ON A HOT STOVE FOR TWO SECONDS THEN IT WOULD SEEM LIKE TWO HOURS.

HOLDS UP FREED OF CITY TO EINST

Alderman Falconer Block to Grant Official Hor to Two Scientists

NEVER HEARD OF HIST

Alderman Friedman Sha in Face of Opponen Calls Action an Ins

time upon time and has beeuuthfully said that the course, if any one swton-Soldner formula tions or ion of starlight near

DOES A TRIVIAL ARITHMETICAL for Einstein. This omission of a simple numerical factor, vi Einstein's calculations of the sun's gravi on the light ray fro said to substa confirm that omewhat sur a public add ny of Scien te triumph iversal grav professor vy, and gov nd. Profes ned the he says

ed deflection, .84 sero ne of the 'Curvature of Space, ous. calculation by an erroneous differenti ng of the path of the ligh. By carel multiplier, two, for the angle, and hene en he integrated the expression to fini s calculation of the bending of the ray, ets only half of the value required by ct physical mathematics. "This error of Einstein was so secu idden that it escaped all the mathematic of Europe for the past thirteen years, a a great mass of literature on relativity exists, a very competent authority has a me that the error might have escaped her century but for the pecul another s which made the discove mstance Island, California, on ble at Mare er 12, 1924. "I set about constru he bending of the wave he theory of relativity, a nade a suitable and accura able to show by simple geom in Einstein's original equation of 1911, which has since continued to be repeated by Richardson, Eddington and the other au thorities on relativity, not one of them sus pecting that all their work was vitiated by an error which a high-school student can now understand. "As an outcome of this discovery, we nov have a correct method of calculating th

IS THE EINSTEIN T. VAGA

IT SURELY IS, according to U. S. Navy, one of the astro bow the knee to Relativity. tions agree not only with Einstein, h formulas deduced over a century ag theory of gravitation fails to explain as above stated, "a crazy vagary" Professor See has himself promulga which he says is rsions; wl n to th in a lett w York can not b ion of the or the ref is a trium dates from andfather h from Be dmitted to and, that i is concerne

rously the unauthorized and indefen
that near the
difficul Capt. T. J. J. See, of the vity?
ed. It omers who has refused to ve de
not reg The recent eclipse observa-
ed, and e says, but with astronomical
mit the o. Furthermore, Einstein's
inent Eu

EINSTEIN'S
ARITHMETIC

owe a duty of truth to the public should
vigorously the unauthorized and indefe
the observed refraction of starlight near th
of the discredited doctrine of relativity?

"3. After ample reflection I have de
'discredited,' and I will now r

"(a) Einstein' at used in 1801,
law of wh numerical value, .84 second of ar
ubled his value, making it 1.68 or 1.7 second
upon the earth, yet he could explain only ha
the Newtonian theory of gravitation. For h
alculation by which, when the ray had passed
arth, he still got only .84 second of arc.

"WORTHLESS, MISLEADING"

Is Einstein's theory of relativity,
says Captain T. J. J. See, because
it is all based on a mathematical
error "which a high school student
can now understand."

T SURELY IS, according to it gra
U. S. Navy, one of the astron ter b
bow the knee to Relativity. ction
ions agree not only with Einstein, t doub
ormulas deduced over a centur e sun
theory of gravitation fails to e.he slightl
as above stated, "a crazy va of be ob

IS THE EINSTEIN THEORY

S ARITHMETIC OFF?

---

and General Theories of Relativity may be said to have shown us a different universe. But their huge scale, their remoteness from the life of man in this world, cannot be denied.

The other breakthrough was into a realm from which you and I and my table vanish at once, simply because the scale is now so minute. The medieval angels dancing on the point of a needle would be vast clumsy creatures in this sub-atomic world. (Incidentally, why hasn't some imaginative physicist tried his hand at a new Gulliver's Travels?) Once we are among the particles and trying to understand quantum mechanics, anything may happen—"waves propagated faster than the velocity of light, inter-change of the time sequence of cause and effect," positrons discovered to be electrons going back-ward in time, particles behaving as they please and so having to be treated on a statistical basis. (Unlike man, who is in danger of losing his free-dom by being regimented on a statistical basis.)

However, the point I want to make here is simply the remoteness of all this from ourselves, occupying as we do a kind of middle position between the stars and these subatomic particles. A great deal has been happening to Time, we can say, but on extreme scales outside our own.

We might compare our traditional idea of Time to a rope that seems whole and unaltered as it passes through our world but is now frayed and insecure at each end far away from us. But if it will no longer do for stars and atoms, it is still good enough for man. In other words, I believe that many scientists (perhaps most of them) cling all the more tenaciously to the classical concept of Time, so far as our life in this world is concerned, because it has been so successfully challenged in its application elsewhere, on the largest and smallest scales. They feel strongly that nobody must start picking at the middle of the rope, where it seems still to hold. This ex-plains—though in my opinion it does not justify —the shouts of "Bosh! Not tested! Not proved! Mere coincidence!" that greet any appearance of parapsychology and ESP, precognition and new theories of Time. I can imagine them crying to each other, with Yeats: *Things fall apart; the centre cannot hold.*

97

It may be thought that I am exaggerating. But at least it is a fact that some fundamental experiments have revealed marked inadequacies in the very theories that are held so rigidly and dogmatically. And when quantum physics come up with queer Time effects or an apparent reversal of cause and effect, many desperate attempts are made to patch up the classical concepts. I promised a brief respite from quotations, and, keeping that promise, I will refer the curious reader, in search of detailed accounts of these matters, to Professor Werner Heisenberg's *Physics and Philosophy: the Revolution in Modern Physics*, and Dr. Mary B. Hesse's *Forces and Fields: A Study of Action at a Distance*.

And now, about to return to quotation, I must first explain how the quotations that follow in the next section were obtained. Knowing only too well that I was not the man to undertake this task, needing somebody much younger, brighter, and far more knowledgeable in science than I am, I engaged an emissary to write to, and then if possible personally interview, a number of distinguished physicists, mathematicians, and philosophers. They were asked for their comments on possible new theories of Time, emerging in the light of precognition and the rest. I have decided rather reluctantly to suppress all names, simply because one or two insisted upon it; and to give some names and suppress others might be thought invidious. I can assure the reader, however, that all the men who were questioned are alike in being undoubtedly distinguished in their particular fields, where they are, to use a fashionable vulgarism, "top people." And even though, in some instances, they may be only "filling our belly with the east wind," I take this opportunity of thanking them all for their various degrees of co-operation.

4.

Here is the first out of the bag, coming from one of the younger men:

It is tacitly assumed in all existing theories in physics that events are simply ordered in time if there exists any causal relation between them, except for certain phenomena which only become important where very high energies or very high velocities are involved. This is why there is such a strong resistance to contemplating precognitive phenomena.

The physics of elementary particles is at present dominated by a clash between the familiar mechanics of objects moving in space and time and certain abstract formulations which attempt to give the essentials of our experimental knowledge about particle processes (disintegration, decay, etc.).

The difficulties raised by the abstract formulations are due to the fact that they involve such bizarre expedients as crossing symmetry, in which the time axis and one space axis are interchanged and in which a particle thereby caused to travel backwards in time is put right by turning it into its corresponding antiparticle going the right way.

And this, no doubt, is a neat trick, but it is not one likely to be much used by people who happen to have had a most convincing precognitive experience, and are beginning to wonder about Time. Let us try another, a very high-level personage, who makes two blunt points:

(1) I am extremely sceptical about the possibilities of precognition and would certainly require some evidence before beginning to take it seriously.

(2) I do not think that the development of quantum mechanics has engendered a different attitude towards the future and its predictability. On the other hand, it is, of course, possible to make fairly good guesses about what is likely to happen in the future in certain cases, but this is rather far away from precognition in dreams.

It is possible, I think, to *feel* the resistance there, and I must add—what I hope to demonstrate in Part Three—that there is in fact a great deal of evidence in favor of precognition, though of course it may not be the kind of evidence this distinguished physicist requires. But an experience remains an experience, even if, because of its peculiar nature, it cannot be submitted to the tests the scientist demands. The sight of Polly Brown may fill Joe Smith with ecstasy, but who will take Joe into the lab to prove it?

I am sorry that my self-denying ordinance prevents my giving the name of the senior mathematician, of the highest repute, who is the author of the following brief statement:

I would emphasize that our present theory is only transitional. Any breakthrough in fundamental physical theory must involve completely unexpected ideas.

It looks as though departures from orthodox ideas of causality in the mathematical formalism will lead to inconsistencies in the interpretation unless these departures are concerned only with extremely short time intervals.

Welcoming this statement, I can only add that it seems a pity, if their present theory is only transitional, that so many scientists should hold it so stiffly—unless, of course, they are aware of its fragility.

Another senior man of considerable reputation and weight takes us a little further:

I think the study of the phenomena of ostensible precognition, and the analysis of the concept of time in current fundamental physics, can be illuminating in that the latter can throw light on the former. Specifically, the *reversibility* of the direction of the spatialised time of statistical microphysics (both classical and quantum), contrasted with the *irreversibility of* the direction of time inherent in spontaneous emission of radiation and radioactive decay processes, suggests a two-dimensional theory of time in physics which offers a logically satisfactory basis for ostensible precognition. The matter is largely a technical one which cannot be summarised in a few sentences but which I am elaborating in a forthcoming book.

And though I doubt if I shall be able to understand the book, I promise here and now to try.

The following are notes taken during a discussion with a high-ranking professor of physics, and if he should see them and find them rather rough, he must accept my apology:

The theory of relativity, and particularly Minkowski's contribution to it, led people to adopt a certain attitude to time. (Minkowski said something like "From now on space and time will lose their independence, and only their union will be real.") This tended to encourage people to think of space-time as ultimate reality which we experience section by section. This has some resemblance to Dunne's theory, but the latter assumes that all events, both past and future, are *there*, whereas Einstein's theory gives no allowance for any mechanism for reaching into the future. Past and future are very clearly separated in the relativistic models—in a sense more clearly than in previous views.

Before Einstein, it was thought that given any event, everything else could be regarded as either *before* it or *after* it. In relativity, some events are definitely

before, some definitely after, and there are some for which it depends on the frame of reference whether they can be called before or after. So the separation is if anything more definite, as this third class falls between the definitely future events and the definitely past events.

Here I must gate-crash this interview to declare, with all due respect, that I do not find this line of reasoning easy to follow. If some men are sober, and some men are drunk, and now there is a third class of men who might be considered drunk or sober, according to a certain frame of reference, I cannot see that this makes the separation of the drunk and the sober if anything more definite. But I must allow the professor to continue:

The development of quantum mechanics has led to a rather different attitude from that encouraged by relativity. In this view the future is unpredictable except in broad outline. Predictions made about averages are always right, but we cannot predict what happens to an individual atom. Einstein did not believe in quantum theory, and commented "God doesn't play dice with the universe."

He ended his talk by quoting a humorous remark, made to him by a German colleague, for which I am particularly grateful: *Wahrheit ist komplementar zu Klarheit.* This I take to mean that truth can be obtained only at the expense of precision, which is something that those of us who are called "woolly-minded" have been aware of for some years. It seems to us that we are living in an age of transition, when it is far less important to clarify such knowledge as we have than it is to reach out for something new and better—before it is too late.

Here is another statement, from a philosopher of science:

In 150 words, one can only be dogmatic, and negative. In my view, there is no *direct* connection between (a) the status of Time in modern physics and (b) questions about precognition and parapsychology. There may be an indirect (sociological) connection: in that the more open-minded the community of physicists becomes about (a), the less rooted may their opposition be to all thought about (b). As to the question, how much one should defer to physics in thinking about parapsychology: remember how disastrously wrong Kelvin was in ruling out Darwinian evolution for thermodynamical reasons. Of course, he could not foresee the 'thermonuclear' explanation of the Sun's continued heat-source, but he was at fault in trusting the *general* correctness of classical physics too much, and the *special* arguments directly bearing on the Darwinian problem too little. (See, e.g., Eiseley, *Darwin's Century*, Chapter IX.) If parapsychological things *are* shown to happen, physics must eventually come to terms with the fact, not the fact be denied for abstract physical reasons. . . .

And with this last remark I for one must heartily agree; though of course if a man, with his theory at stake, makes up his mind not to *be* shown that queer things are happening, little can be done except to make a face at him and a rude noise. Though a little tempted, I shall not do this to the next professor, part of whose letter to my emissary I quote:

I do not consider that theories of Time could have any useful part to play in this. Nature is a flux to us, and this flux *we* atomise into isolated events. To compensate for the isolation *we* introduce Causality and Determinism. How far this is itself demanded by Nature as a necessary part of our thinking, I cannot say. Causality and Determinism require that the isolated events be arranged or ordered in a rational pattern, and our ordinary mundane macroscopic experience enables us to do this ordering via two abstractions of the events. So we do ordering in *space* and ordering in *time*. Thereafter to take "Time" as something in itself, as a self-subsistent entity analogous to the Matter-Space-Time events from which Time has been abstracted, is a piece of mental acrobatics. If it is likely to be now suggested that this "Time" has some special properties that make possible parapsychological effects of such a nature that events not yet capable of being named or even "realistically" thought of (because they are not yet part of the flux) can be forecast—then all I can say is "Produce for me a sample of Time as an event."

Yes, my dear Professor, but you must try to bear with us a little—that is, those of us who believe we have had certain experiences that cannot be fitted into your pattern. If we regard Time as one of the conditions of human experience, then we may have to take a new view of it (those of us who do not enjoy being regarded as boobies, self-deceivers, charlatans) to account for these other kinds of experience. Of course, I agree with you when you say, in a later part of your letter (a part I have not quoted), that "one has no way of statistically assessing the chances that an apparently precognitive dream may be no more than coincidence." But let us take it easy here and not wall ourselves in with the statistical method.

"I venture to maintain," another professor has declared, "that scientific progress has never yet resulted from simply dismissing incongruous phenomena as negligible illusions just because they refuse to conform."

Finally—and it is a pleasure to welcome him here—the following is a statement from a distinguished astronomer, many of whose timely public protests, on other questions, I have applauded:

Firstly, I think that the problem of time in modern physics and astronomy is an issue of great importance. The fundamental issues which determine the direction of the arrow of time are not at all clear, and there is beginning to be a good deal of discussion about this matter. . . . As far as my own personal situation is concerned the problem is one in which I am intensely interested, not only because of its inherent relevance to day-to-day problems in astronomy with which we deal, but also because from a personal point of view I have had a number of dream experiences during my life which appeared to have conveyed information, not so much about future events, but about events which were in process of taking place and with which I subsequently became acquainted a short time afterwards. In general, although the nature of my training should persuade me otherwise, I am nevertheless forced to suspect that certain processes are operative which are not included in the contemporary laws of physics. . . .

Well, there it is; I have emptied the bag; and I think it will be agreed that on this level—and it is a top level—the attitudes of mind range from those expressing extreme resistance to any possible challenge to conventional theory, to those expressing a skepticism about "the contemporary laws of physics" and an open mind about possible future developments. And my guess is that if this inquiry had been pursued on a lower level, where dogmatism reigns, there would have been a great deal more resistance, and the answers would have been shorter, blunter, and more foolish.

All I wish to do in this section is to offer some proof, by way of references and quotations, of my remark in the last chapter that the Time situation in contemporary scientific and philosophical thought has eased a little. If I am accused of merely dipping here, snatching at gleaming morsels like a bird at a pond, I will not plead complete innocence, though I have in fact read a great deal more than will appear and can justly claim that my space is limited.

A first dip, then, into K. Gödel's *Relationship Between Relativity Theory and Idealistic Philosophy*:

One of the most interesting aspects of relativity theory for the philosophical-minded consists in the fact that it gave new and surprising insights into the nature of time, of that mysterious and self-contradictory being which, on the other hand, seems to form the basis of the world's and our own existence. The very starting point of special relativity theory consists in the discovery of a new and very astonishing property of time, namely the relativity of simultaneity, which to a large extent implies that of succession. The assertion that the events *A* and *B* are simultaneous (and, for a large class of pairs of events, also the assertion that *A* happened before *B*) loses its objective meaning, in so far as another observer, with the same claim to correctness, can assert that *A* and *B* are not simultaneous (or that *B* happened before *A*).

Following up the consequences of this strange state of affairs one is led to conclusions about the nature of time which are very far reaching indeed. In short, it seems that one obtains an unequivocal proof for the view of those philosophers who, like Parmenides, Kant, and the modern idealists, deny the objectivity of change and consider change as an illusion or an appearance due to our special mode of perception. The argument runs as follows: Change becomes possible only through the lapse of time. The existence of an objective lapse of time, however, means (or, at least, is equivalent to the fact) that reality consists of an infinity of layers of "now" which come into existence successively. But, if simultaneity is something relative in the sense just explained, reality cannot be split up into such layers in an objectively determined way. Each observer has his own set of "nows," and none of these various systems of layers can claim the prerogative of representing the objective lapse of time.

It is only fair to the author to add that he then presents arguments that challenge this inference.

One of the most thorough-going and rewarding works in this field is M. Johnson's *Time, Knowledge and the Nebulae*. After marking many passages for possible quotation, I now find I have space only for one, and the choice is difficult. But here at least is a tasting sample:

We all recognize that a foundation of human character is to learn to build into our personality a shrewd selection from the glory and agony of passing events, so that experience of the most transient can contribute to something of us which is not so transient, and may even reconstruct our mental structure through active memory and imagination. The "passing" is not necessarily the "lost". Perhaps this is one sense in which we train ourselves to conquer the ancient enmity of time. But it is not a sense susceptible of scientific—even contemporary psychological—analysis; the concepts involved imply judgments of Value and not merely of Fact, and lie entirely outside the dictionary of the physicist. Not even the psychologist has found adequate words to encompass more than their significance to the external spectator. Perhaps the final word lies with the metaphysician, but perhaps only when he understands the artist or poet as well as the logician and physicist.

And perhaps I chose that particular passage just to be able to exchange a passing wink with the artist or poet.

Reading the late Hans Reichenbach, I noticed in him a pleasant trait, not absent from some of his colleagues and not without significance. He was a sufficiently rigorous philosopher of science ("There is no other way," he wrote, "to solve the problem of time than the way through physics"); but he could not resist describing what would happen if, for example, we occupied a torus space, or a man had a world-line that went off into a loop, which, if you were that man, would bring you the following experience:

Some day you meet a man who claims that you are his earlier self. He can give you complete information about your present condition and might even tell you precisely what you are thinking. He also predicts your distant future, in which you will some day be in his position and meet your earlier self. Of course you would think the man insane and would walk on. Your companion on world-line 1 (normal world-line) agrees with you. The stranger goes his way with a knowing smile; you lose sight of him as well as of your companion on world-line 1 and forget about both of them.

Years later you meet a younger man whom you suddenly recognise as your earlier self. You tell him verbatim what the older man had told you; he doesn't believe you and thinks you are insane. This time you are the one that leaves with a knowing smile. You also see your former companion again, exactly as old as he was when you last saw him. However, he denies any acquaintance with you and agrees with your younger self that you must be insane. After this encounter, however, you walk along with him. Your younger self disappears from sight and from then on you lead a normal life.

Because of course you have now got out of the loop in your world-line. But the point I wish to make about this not uncommon playful kind of example, not necessary to the main line of argument, a sort of self-indulgence, is that it seems to me to represent a protest of the imagination against theories that are too narrow and held too dogmatically. And if I found myself compelled to indulge in such fantastic examples, enjoying every opportunity to present them, I would begin to wonder if my intellectual loyalties were not keeping too close a hold upon my whole self, if indeed the theories I held were good enough.

Returning to more exact thought, I have received from my friend Professor Fred Hoyle what amounts almost to a considerable paper, though not intended for print, on *A Physicist's View of Time*. (It is in fact a considered reply to a question I asked in conversation.) It is too long to be included here, and I hope Professor Hoyle will forgive me—and his colleagues please refrain from criticism until the completed paper appears in print—if I give only his summary:

The location of physical events in the world require four numbers, three for space, one for time. The events that constitute a particle form a line. If matter is created, the line has one termination but not two. Proceeding along the line towards the termination is defined as going into the "past," proceeding in the opposite direction represents going into the future.

Particles interact with each other in such a way that physical fields, the vehicles of communication, are represented by half-cones, not by full cones; when a

particle radiates, the radiation propagates along the half-cone that contains the future section of the line of the particle (Fig. 1). This asymmetry produces the one-way step ladder system of communication (Fig. 2). It leads to the causal character of the Universe. In particular the realm of experience described by thermo-dynamics is a consequence of this asymmetry. Thermo-dynamics has no deeper significance.

It is imperative in the asymmetric system that the radiation be considered as a grand total, not only of the direct effect of the particular particle itself but of all the reflections from all other particles. It is possible that reflections come from the future. Indeed, it is possible to take the view that if it were not for reflections from the future the everyday world of common-sense would not exist.

In an early section of his argument, on *Personal time as an absolute or invariant quantity*, Professor Hoyle remarks rather casually: "Although a human being is really a complex collection of particles it is possible to think of ourselves, for ordinary macroscopic purposes, as single particles." Well, this complex collection of particles thanks him, and apologizes again for making such poor use of his paper, for doing justice neither to science nor to friendship.

Still pecking around and dipping, I have found an article in the Journal of the Franklin Institute, by Dr. Gustav Stromberg, on *Space, Time and Eternity*. After offering some evidence for a five-dimensional universe, which would include a realm beyond the four-dimensional space-time world of physics, an "eternity domain," he proceeds to observe:

Certain properties of the eternity domain can be deduced from the fact that it lies beyond both space and time in their physical sense. No physical measurements can be made in that world, because measure-

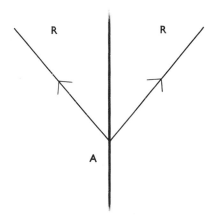

Above, a diagram illustrates Professor Hoyle's view of Time as an asymmetric system. The center line (A) is the chain of events that constitute a "particle" and its duration in Time; when a particle radiates (that is acts upon its surroundings), its influence (R) is propagated within a half or asymmetric cone (shown here in section) that contains the particle's future line. The diagram below shows how this asymmetry results in a "one way step-ladder system of communication" between any two particles (A and B). According to Professor Hoyle, this system of "interaction" leads to the causal character of the universe.

Left, drawings of the 19th-century Danish philosopher Kierkegaard (and a friend) illustrate Professor Reichenbach's anecdote about a loop in a man's "world-line." While out with a friend, you meet a man who claims to be your older self. You walk on, losing sight of the stranger and of your companion. Years later, you meet a young man whom you hail as your younger self. You also recognize his companion as the friend of your youth—no older than before, since you are at the point where the loop began. Having returned to your world-line, you walk on with your companion and lose sight of your young self.

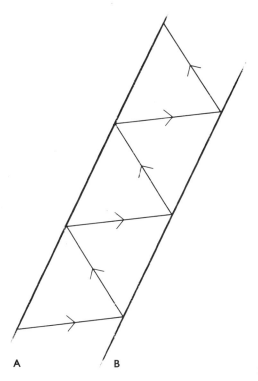

A   B

ments are made with material measuring rods and clocks, and matter has no place in a non-physical world. Since there can be no units of length or of time in the eternity domain, it must be described as *non-metrical*. Separations in space cannot be defined, although there may be some kind of "mental separation". Long and short time intervals cannot be distinguished from one another, so that *duration* in time is meaningless. On the other hand, there are reasons to believe that a *sequence* in time has a well-defined meaning. We may expect that there is a causal relationship between events in the physical and the non-physical worlds. . . .

And just as a parting shot, though I am not sure at whom, I add a short passage from a broadcast talk given by the distinguished Oxford nuclear physicist D. H. Wilkinson. It refers to the possibilities opened up by the existence of such subatomic particles as the neutrino, which interacts only in a very weak fashion with other particles and is extremely unlikely to interact with any particle at all in passing through the earth. It is just possible, then, that there might be independent but interacting (however feebly) universes:

Perhaps there do exist universes interpenetrating with ours; perhaps of a high complexity; perhaps containing their own forms of awareness; constructed out of other particles and other interactions than those which we now know, but awaiting discovery through some common but elusive interaction that we have yet to spot. It is not the physicist's job to make this sort of speculation, but today, when we are so much less sure of the natural world than we were two decades ago, he can at least license it.

Certainly; and all the particles here (perhaps a neutrino or two have somehow crept in), always appreciative of sense and courage and the open outlook, are giving that last statement a cheer.

6.

Though the scientists have quit the scene now, I am not closing this chapter yet. I am keeping the promise I made in the last chapter: to put up a defense against the charge, a grave crime in metaphysics, of "spatializing" Time. When we do this (no doubt for shady unphilosophical purposes of our own) we confuse the spatial relations between things with the temporal relations between things. We need the concepts of Space and Time to help us to "sort things out"; without them we would never be able to make head or tail of anything. If there were no things, there would be no Space; if there were no changes in things, there would be no Time. (Neither of them should be thought of as a container, though we all do it.)

The fashionable philosophical charge against lay time theorists is that we confuse spatial and temporal relationships, and that we imagine the latter to be between things and not "events." Because of this further confusion, we are really assuming that changes can change, which is nonsense and ready to become the father of any kind of conclusion we prefer. The argument is a highly technical one, and I have no space here, even if I had the competence, in which to refute it. All I can do is to make several points in our defense that seem to me generally important.

I will admit at once that my thought refuses to dwell where Space and Time can be entirely separated. If I stay there I cannot think to any purpose about either of them. (But I promise to spot the flaw in Dunne's technical argument, leading him into an infinite regress, when I come to it in Part Three.) I do not think I am peculiar in this. Everybody instinctively "spatializes" thought about Time, even the philosophers when they are off duty and are not fully armed to protect scientific conclusions. That is the way our minds run, and there is no help for it; when we are thinking and talking easily and freely, not toiling in the mines of logic and linguistics, we keep on turning Time into Space, passing the cake of reality from one sister relationship to the other, who seems to have the larger appetite.

The scientists do it. No doubt if pressed they would declare that "dimension" in their references to a "fourth dimension" is used only in a special sense, which we laymen do not understand and therefore illegitimately enlarge. However, I have made my way lately through too many scientific treatises not to observe that, when their authors are well in their stride and happy, they "spatialize" as much as the rest of us do. What is "a world line" if not something thought of spatially? Take the "spatialization" out of all these accounts of the space-time continuum, restore the time order there to events, and their authors would no longer be happily in their stride. They are obeying an instinctive response. After all, they live here with us, not on the moon.

I maintain we cannot help turning Time into Space, always beginning again to see temporal relations as spatial relations. I say that our critics, when not engaged in their professional exercises, do it just as we do. All of us, when asked to consider a period or a world figure, let us say the 19th century or Napoleon, do not see it or him as a succession of events but as if spread out in a spatial order. If our use of these concepts of Space and Time is a condition of our thinking, a further condition seems to me that Space keeps on swallowing Time.

It is argued that because we "spatialize" Time, we draw wrong conclusions about reality, as for example that the past still exists. But the people who hold this belief have not arrived at it through bad semantics, logical fallacies, faulty syllogisms; they are trying to understand some experience; and here their philosophical critics offer no help at all.

Unless I am far more eccentric than I imagine myself to be, I am right in also thinking that we all "spatialize" to some extent our own time. At least we do not think of our own time as a succession of events each canceling out the one before it. Certainly I for one do not see myself as what I am now, and nothing else. I seem to stand, as my essential self, outside the time order. Fifty years ago I seemed to myself older than other people assumed me to be; now I seem younger. It may be objected that I am now attempting to oppose a philosophical argument with mere psychological quirks; to which I can only reply that I live with and by the psychological quirks, not the philosophical argument.

If what we are all trying to do is to describe reality, an intuitive interpretation of actual experience, however loose the terms it employs, may bring us closer to it than the most rigorous linguistics and the severest logic. These last may

leave so little room to maneuver that they cannot capture any experience outside an accepted narrow pattern; and indeed they may operate (for philosophers do not live in a vacuum) to insist upon that pattern and to deny anything outside it.

Moreover, in our encounter with the ordinary physical world, our spatial and temporal experiences simply refuse to be as widely separated as the scientific philosopher's Space and Time. Let us say I travel by car, at 60 miles an hour, across the five miles of the Little Puddlefield district. I see a church, two farms, four bungalows, and an inn, successively within five minutes; I have a Time relation with this region. On the next occasion, I fly over it in a jet plane on a clear day, look down and see all at once the church, the two farms, the four bungalows, the inn, and what was in Time is now in Space. The difference in ordinary experience is simply between two modes of travel.

And in each instance, so that the objects could be recognized at all, either in Time or Space, two sets of two-dimensional images have had to be focused and reversed and introduced to the idea of a third dimension and inspected and corrected by all kinds of knowledge about the world. True, we are not aware of all this in ordinary experience; if we were, we should not be able to do anything except lose our sanity. But neither are we aware of any fundamental distinction between Space and Time, any gap never to be closed. Outside highly technical discussion, we are for ever abolishing that distinction, closing that gap, not only for all practical purposes but in a sincere attempt (however unphilosophical it may appear to be) to understand reality.

When I write as if Time could have dimensions, I realize that I offer myself to philosophical chastisement. But groping within that *as if*, I may come near to grasping a truth that altogether eludes all linguistic refinements and the machinery of logic-chopping. It is better to risk confusing Space and Time and move on than to sit at the bottom of a chasm between them, getting nowhere, unconsciously defending a wrong idea of man's life.

4   Time in Fiction and Drama

In the German film *Waxworks*
(1929), the crazed tsar Ivan the
Terrible sits frenziedly reversing a
huge hourglass, for he has
identified the glass's trickling sand
with the duration of his own life.
Though views or theories of Time
are seldom the *subjects* of fictional
or dramatic works, a writer's
treatment of Time is always an
essential element of his technique.
Also, in a play or story (as here) a
particular character's awareness of
Time may often be a means of
heightening dramatic effect.

4 Time in Fiction and Drama

I.

Our modern society, especially when it is urban and industrial, is of course quite unlike earlier societies, except perhaps that of the later Roman Empire. And it is not only our amazing progress in science, invention, technology, productivity, that sets us apart from earlier men. There are other differences equally important, even though they are rarely mentioned.

We may call our society a Christian one, we may sincerely believe that it is, but now it really has a secular basis; and, whatever we may say, we believe it to be governed not by divine but by political and economic laws. Unlike earlier men, we do not feel that our society is contained by common religious beliefs and by great

myths. Our daily life, as Professor Mircea Eliade has pointed out, is no longer symbolic and sacramental. Ancient traditions no longer have any meaning. Unless we are artists or pursue a definite vocation, we work to earn a living, in the hope of being able to afford some luxuries and to be able to amuse ourselves during our increasing hours of leisure.

Yet our ancestors live on in us, if only dimly and remotely, well below the level of the typical modern consciousness. And it is when we seek escape and some refreshment of the spirit in our distractions and amusements (and we have far more of these than any earlier society had, just because we need them more) that we join hands with these remote ancestors of ours. There is a difference, of course: What for them was public, communal, entirely serious, has dwindled for us into some private, entirely individual way of merely amusing ourselves and, as we say, "passing the time." This is, however, truer of the middle-aged and elderly among us than it is of the young, who may be nearer in spirit to their ancestors or may be rebellious because they feel instinctively that modern society will deny them too many primary satisfactions.

All this helps to explain the astonishing popularity of the heroes of detective fiction and "westerns," stage and film stars, "television personalities," remarkable athletes and players of games. They seem to exist in all that is left us of the old mythological atmosphere. When the

Left, Mexico's Emiliano Zapata, leader of the "Death Legion"—a horde of lawless bandits dedicated to the return of Spanish-held land to the Indians. After Zapata's murder in 1919, many of the Indians asserted that their leader, like some mythological hero, still lived and would someday return to lead them again.

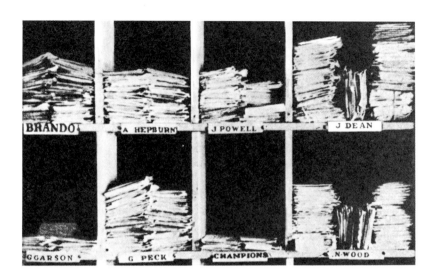

Left, the American actor James Dean as he appeared in *Giant*—his last film role before his death in 1955 at the age of 24. Above, some of the mail that accumulated at the studio after Dean's death. It was written by fans who refused to acknowledge that their idol was dead, and who (like the students below) continued to collect his photographs and run his fan clubs.

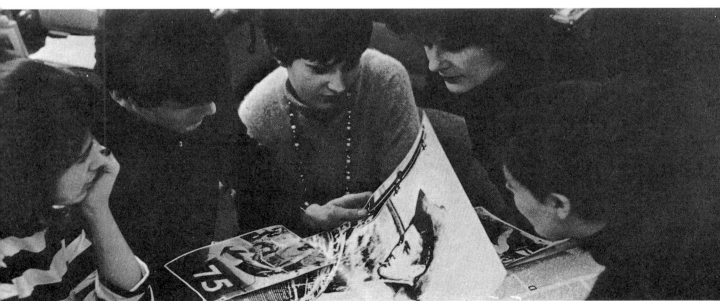

young offer these personages an apparently hysterical adulation, they are attempting blindly to restore to our society a symbolic, mythological, sacramental element it has lost. (The young film actor James Dean was killed in a car crash, but numbers of young Americans refused to believe this, because to them he was an immortal.) Elderly critics of teenage behavior should save some of their time and energy for a critical examination of our society, asking themselves how it responds to the instinctive high expectations of adolescence. Rebellious boys and girls may behave foolishly, but I suspect they cannot help feeling that as they grow up they find themselves becoming enclosed between huge and mysterious barriers.

And one of these has been created by our idea of Time. It is no accident that one of the main themes of popular science fiction is "time traveling." To hop about from age to age, jumping thousands of years, is an easy and obvious way to defy Time. But the millions of paperback books, all the stages, and the film and television screens also offer some means of escape from relentless chronology, from the ticking away of youth and health and high spirits toward senility and extinction.

We must now move rather carefully. It seems to me that what chiefly helps to provide us with rest and refreshment in fiction or drama (over and above, of course, our interest in the characters and their adventures) is simply a different kind of duration or a changed temporal rhythm. We must not assume, for example, that an historical novel or play, or one covering an unusual length of time, has any special merit. Works of this sort may have a particular appeal to certain readers or playgoers, but not, in my opinion, for the reason I have suggested—not in terms of Time.

Few readers can be more aware of the Time element in fiction than I am, but this does not mean that I am prejudiced in favor of cradle-to-the-grave novels, in which we attend the birth of the hero or heroine in chapter one and his or her funeral 50 chapters later. Indeed, I usually dislike novels of this kind, and thoroughly detest them when they fasten on to some historical figure, offering me bogus biography, an unpleasant mixture of fact and fiction.

No, the amount of time taken by the action of a novel is of small importance. It is the way in which that time is treated that is important, giving us, as readers, a different kind of duration, a changed temporal rhythm. There are three main ways or methods.

The first method belongs to the born novelist, the man or woman God intended to write novels. It is, I believe, almost entirely instinctive. An even flow of narration is maintained; a whole wide and rich scene is kept on the move without jerks and breaks; and although the days and weeks, months and years, have been enormously speeded up, they seem to pass smoothly at an even rate. It is narrative moving in top gear. Tolstoi is a master of this method; but there are lesser novelists who, whatever else they lacked, had this particular rare gift—Thackeray, for example. All novelists who are successful with this method make us feel we are enjoying a marvelous memory. We are in a time not unlike our own but cunningly accelerated, observing the world as a demigod might observe it. The release from our own kind of duration is obvious.

The second method may be found in highly dramatic novelists like Dickens and Dostoevski. They give their time impatient jerks and then suddenly slow it up, to create tremendous scenes, almost as if they were about to write for the theatre. By being so masterful with their times, now hurrying everything along relentlessly, now slowing up to make a scene, as if they were driving a car that went at 80 miles an hour and then at 10, they play on our nerves, compel us to feel excitement, often strange forebodings. At one moment these novelists seem unreal, then at another moment they seem more real than life itself, showing us what it is *really like*. (Readers who feel this is not true of Dickens should read his later novels with more critical attention.) Such novelists are of course unusually gifted in many different ways, but the Time element is important in their work. They are impatient, ruthless *masters* of Time, imposing upon us unfamiliar temporal rhythms. As we surrender to them, it is as if we had a heightened temperature.

Left, the castle of the sleeping beauty—an illustration by France's Gustave Doré of the story of a princess who falls asleep for 100 years. In this story, as in many such tales, the treatment of Time and duration adds to the ''unreality'' of the fairy-tale world.

Below, the great Russian novelist Leo Tolstoi telling stories to his grandchildren while waiting for a train. Tolstoi's novels cover vast periods of time; yet, throughout them, events move at a smooth and constant pace that conceals from the reader the acceleration given to the passage of time.

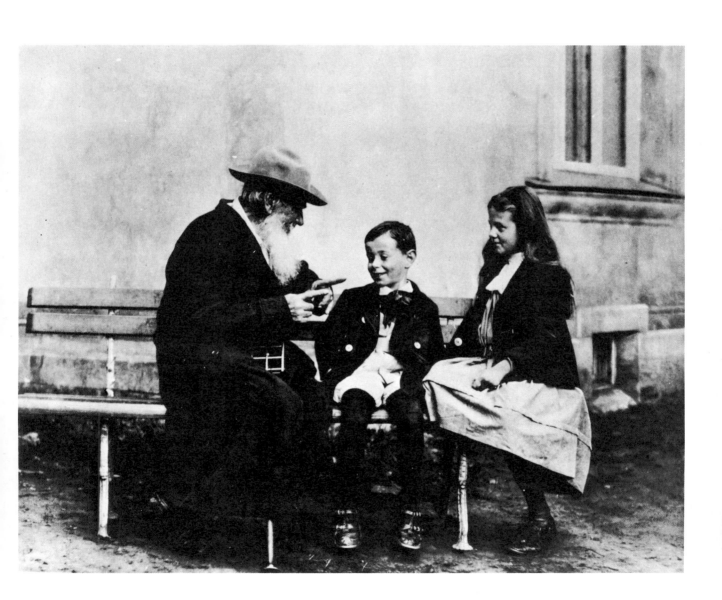

Left, Napoleon's retreat from Moscow (1812) by an artist of the period—one of the climactic incidents in *War and Peace*, Tolstoi's epic novel of aristocratic Russian society.

Below left, from the 1946 film of Dickens's novel *Great Expectations*, Miss Havisham shows Pip the table laid for her wedding banquet many years before. In this room the past seems imprisoned among the dust and cobwebs—an impression of arrested Time that underlines the old lady's bitterness and grief.

Three modern novelists whose work reflects a distinctive treatment of Time: Right, France's Marcel Proust, whose long novel *A la Recherche du Temps Perdu* is a minutely detailed recreation of the past. Below, the German novelist Thomas Mann, who (in *The Magic Mountain*) contrasts "real" Time with the accelerated time scale of a dream. Below right, England's Virginia Woolf, whose characters seem to move in an intense, complex world remote from conventional operations of Time.

The third method could be described as that of slow motion. Its account of some brief episode may take us longer to read than it would take us to experience in real life. Moments are expanded so that nothing they contain is left unnoticed and unrecorded. It is as if a supremely intelligent ant, well acquainted with human life, began to report our talk, minutest actions, gestures, grimaces, and swiftest glances. Sterne, to whom modern literature owes much, was an early and triumphant master of this method. Sometimes in *Tristram Shandy* we feel that everything is coming to a standstill, that all temporal flow has ceased, that the hands of clocks must move through glue, that the next minute had been turned into a gigantic obstacle.

And of course many novelists of this century have made use of this slow-motion effect; the "stream of consciousness" kind of narration would be impossible without it; and once again the escape from our familiar duration is obvious. We are magically transported into another and very different time, one that is not only moving very slowly but is also small-scale and very private, far removed from public events and history. This slow-motion and intensely private time can be used to create a certain strange beauty, as in some of Virginia Woolf's novels, or a memorable comic realism, as in Joyce's *Ulysses*. And one reason why it seems to represent our age is that it is a revolt against the tyranny of passing time.

Below, an illustration by Hogarth to Volume II of *Tristram Shandy* depicts Corporal Trim reading a sermon to Tristram's father and uncle and their friend, Dr. Slop. Despite its title, the novel concerns the "life and opinions" of the father and uncle rather than those of Tristram himself, who is not born until Volume IV.

TRISTRAM SHANDY. VOL.II.Ch.6.P.12.
Corporal Trim, reading the Sermon to
Shandy's Father, Dr Slop and Uncle Toby.

Left, a caricature of the 18th-century English novelist Laurence Sterne rejecting the summons of death—an illustration of Sterne's temporary recovery from a bout of consumption. This episode is described in *The Life and Opinions of Tristram Shandy*—Sterne's nine-volume novel that contains so many digressions that it often seems that the "temporal flow" is at a standstill.

So much for fiction; now for drama. We must consider drama, of course, in terms of its performance and our relation to it not as readers but as audience. When we attend an adequate production of a good play, we enjoy an experience that is quite different from anything we know as readers. In *The Art of the Dramatist*, I maintain that this experience, which I call "dramatic experience," is unique. Fully to appreciate a play we have to maintain a delicate balance between what is taking place apparently on two different levels of the mind. On one level we are involved in the drama, are living imaginatively with its characters. On the other level we are enjoying a performance by actors on a stage, being fully aware that we are in a theatre.

If we are too much involved in the drama, like the mining-camp audiences who would intervene to save the heroine from the villain, or if we are aware of the actors as actors and of nothing else, like certain sections of audiences that are there to worship some star performer, we destroy this balance and so do not achieve unique "dramatic experience." In the same way, the dramatist has a double task, creating the imaginary life of his characters and at the same time writing for a certain number of players on a certain kind of stage. It is this mixture of sheer creation and a highly skilled technique that makes playwriting difficult. Even in plays considered worth a production, nine times out of 10 there is an obvious deficiency on one side or the other.

What has all this to do with Time? So far as audiences are concerned, it may help to explain a certain curious indifference to the Time element in drama. Audiences are indifferent because, if my theory is true, their minds are so busy achieving the true "dramatic experience"; and indeed, with a first-class production of a play of supreme merit, their minds may be said to be stretched. (I am not referring now of course to the actual duration of the play's performance. This is something else, and, as everybody who has worked in the theatre knows, it is of very great importance.) When we read a novel, we please ourselves when and how we spend our time with it. But in the theatre we do not attend to Act I on Tuesday, Act II on Wednesday or

In his novel *Ulysses*, James Joyce examines in minute detail the thoughts and actions of a day in the lives of two middle-class Irishmen. Such detailed description of a single day produces an effect of Time passing in "slow motion"— the reverse of the "accelerated" coverage of long periods by writers such as Tolstoi. Above, one of the two men, Leopold Bloom (in a dramatization of the novel produced in London in 1955) relives an episode from his unhappy past.

Thursday, Act III perhaps the following week. We have to sit through the whole performance, which means that toward the end we are probably becoming rather restless, rather irritable, quick to resent anything that appears to offer our fatigued attention rather less than it deserves.

This explains why experienced and highly skilled playwrights and directors artfully quicken the pace in the later scenes. It also explains why there are fewer outstanding plays than novels. Even novelists who have written masterpieces often begin better than they end, whereas the dramatist who writes a poor last act is not forgiven, the audience demanding that their continuous attention should be sufficiently rewarded.

When, however, I suggested that audiences, with their minds busy or even "stretched," are curiously indifferent to the Time element in drama, I was not thinking about the actual duration of the performance. I was referring to the drama's own relation to Time, which is quite different from our own relation to chronological time. And the greater the play, the finer the production of it, the less we concern ourselves with the Time element. Or so it seems to me.

When I am attending a fine production of one of Shakespeare's great tragedies, I have only the haziest notion of the Time element in its huge terrible world. I have little or no *temporal* idea of the drama. Othello and Hamlet, Macbeth and Lear, move inexorably toward disaster and doom; and here they must be in some kind of time, but I am hardly aware of its passing. It is all the same if the action takes place in hours, days, weeks, months, years. Indeed, although I feel I know these tragedies, I would have to go back to the text and then make some calculations to discover the amount of time Shakespeare formally allotted to them. And I do not believe he cared any more than I do. Time is the almost invisible servant, not the master, of this drama.

I may be told that I am forgetting classical drama and the famous unities. But even here, I maintain, the audience is not really concerned

Engravings from Cervantes' *Don Quixote* shows the credulous Don's absorption in a puppet show (left) and thus his loss of the detachment that, as well as involvement, is necessary for the true "dramatic experience." Finally, enraged at the heroine's ill-treatment, he wrecks the whole show (above).

with the Time element. And I feel I have some claim to be heard on this subject, if only because two of my plays, *Dangerous Corner* and *An Inspector Calls*, which have been performed many times in many places, do in fact offer the audience a continuous action (broken into acts only for convenience) in a single setting, and are contained within the unity of time, place, and action. (It is true, however, that most of the action of *Dangerous Corner* takes place in a kind of loop in Time, returning toward the end to a moment at the beginning.)

But the reason for writing plays in this form has nothing to do with the Time element. It is because their action works like a coiled spring, producing an effect both of increasing tension and dramatic inevitability. In plays of this kind (of which perhaps the supreme example is the *Oedipus Rex* of Sophocles) we are made to feel that the characters are helpless victims of fate. But it is this atmosphere and this kind of action that demand the particular closely-knit form. The fact that the time passed in the play may equal its actual duration, the time spent by the

Right, France's Marcel Marceau, in his dramatic version of the seven ages of man, mimes man's progress from birth to death (which are each represented by a variation of the same pose).

Below, a scene from the Italian playwright Pirandello's *Six Characters in Search of an Author* (1921). The play concerns six characters who appear during a rehearsal by a troupe of actors. The characters explain that, though they are the creations of an author's imagination, their story has never been written in the form of a play. Then, watched by the actors, they take over the stage and "act out" the drama of their lives. Throughout, the action operates on two levels of Time: the "real" time of the actors and "illusory" time of the characters. (The actual time of the audience is a third level.)

audience, is of no real importance, in spite of the classical fuss about the unities. The criticism based on them was merely pedantic, not really concerned with actual experience in the theatre.

Those of us who have worked in the theatre know very well how an audience will accept without protest a stage time moving at least twice as fast as their own. For example, a play may be written in a severely naturalistic manner and have a production aiming at an effect of complete realism. Yet during the course of a single act, perhaps lasting 45 minutes, a host and hostess can be shown awaiting their guests, cocktails and then dinner can be served, after-dinner talk over coffee and brandy can be heard, and guests can begin to leave; so that three hours of our ordinary time has been compressed into

45 minutes. And if the playwright, the director, and the players have all done their work properly, the audience will accept this as a strictly realistic presentation of a dinner party.

Some allowance must be made, of course, for the fact (never really lost sight of, if my theory of "dramatic experience" is accepted) that, naturalistic or not naturalistic, realism or no realism, whatever happens on the stage is governed by certain widely accepted conventions. But there is skill, too, in this artful compression of time, in the way in which meaningless moments are carefully excluded and whatever is said and done will be meaningful. Empty time is banished from a good theatre.

We might say, then, that great poetic drama takes us outside Time, while our contemporary

In a scene from the 1955 London production of Samuel Beckett's *Waiting for Godot,* the two tramps who are aimlessly waiting for an enigmatic Godot are interrupted by the overbearing Pozzo and his slave. This strange fantasy world seems to be outside the ordinary passage of time—an impression that emphasizes the apparent futility of the tramps' vigil.

prose drama, if it knows what it is doing, at least offers us some release from our usual sense of passing time by showing us people who are living in a compressed and accelerated time, in which everything seems to us meaningful and significant, so that we observe these people as a demi-god might observe us and our affairs. This is true too of films, which make use of a very flexible medium, capable of playing all manner of Time tricks; but there is one important difference between a film and a play. When I watch a play I feel I am looking at something that is happening: It is all in the present tense. When I watch a film I feel I am looking at something that has already happened: It is all in the past tense.

On the other hand, although television drama seems to have more in common with the film than with the theatre, I feel that it is all in the present and not the past tense; it is happening and has not already happened. (This may be due to the fact that we associate the television screen with events that actually *are* happening, with news items, sports, and so on.) This appears to suggest that television drama can combine the advantages of film and theatre, having the flexibility of one and the immediacy of the other.

And indeed, moving a long way from most of the half-hearted domestic entertainment of today, it is not difficult to imagine enormous television theatres in which audiences are able to watch, projected in full color on an equally enormous screen, performances taking place in the same building, thereby enabling audiences and players to respond to each other as they do in the theatre. Time, working with something rather better than commercial acumen, may yet bring us some such technological marvel, thus helping us to forget it.

2.

Now we come to Time as a *subject* for fiction and drama. When I originally scribbled a brief synopsis of this projected book, I included this topic and gave it no thought, being convinced that no section of the work could be easier to do than this one. As soon as I did give it any thought, I felt, scores of novels and plays having Time as a subject would return at once to my memory. And

I was wrong. (I am often wrong, unlike most of the politicians, editors, and experts I know.) One reason why I went wrong here should be obvious now—it is painfully obvious to me— because it is concerned with the nature of Time.

A novel or a play cannot really be *about* Time. (And I ask the reader to remember that I am a man who is widely credited with having written "Time plays," although I never made any such claim myself.) Time is a concept, a certain condition of experience, a mode of perception, and so forth; and a novel or a play, to be worth calling one, cannot really be about Time but only about the people and things that appear to be *in* Time. Some novelists and dramatists may be unusually aware of Time, but they have to write about something else.

Let us take an example that immediately suggests itself: H. G. Wells's little masterpiece (for that is what it is) *The Time Machine*. If ever there was a Time tale, this is it. But once its introductory and explanatory chapters are out of the way, even *The Time Machine* is not about Time; it is about human society in the far future and finally the last days of the world itself.

Stories of the future, which have recently been piling up as contributions to science fiction, never take Time as their subject. They, too, like *The Time Machine*, are concerned with persons and events remote from us *in Time*. They do not describe the container but what it contains. They may offer us, as we shall see, some hazy or wild accounts of what they like to call "time-traveling"; but most of their authors are not

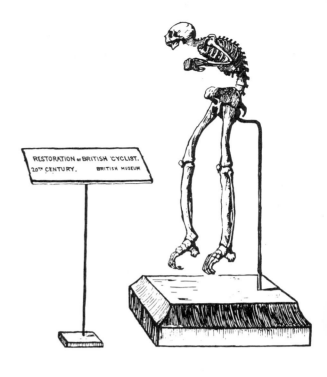

Above, a *Punch* cartoon (1889) of a skeleton of a "20th-century British cyclist"—a satirical comment on the wild ideas about man's evolution that were current at the time. Such speculation about the future development of mankind and of society was often a theme of England's H. G. Wells (1866–1946), seen right, in a photograph taken at the height of his activity as a social reformer during the 1920s.

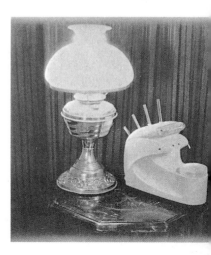

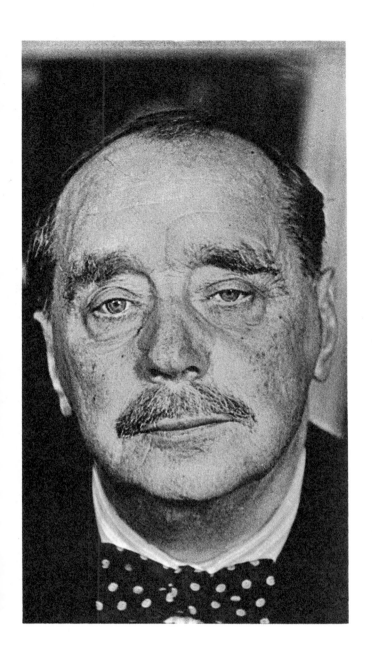

Above, a "Time-cheating" caricature (1929) by Max Beerbohm of H. G. Wells confronted by his younger self. Both Wells's youthful interest in science and his later concern with the structure of society are reflected in *The Time Machine* (1895)—the story of a future society visited by a time traveler. Below, photographs show a model of the time machine taking off on its journey into the future.

thinking about Time at all but about possible developments of our society, planetary or galactic wars, visitors from space, and so on.

Turning over in my mind now the hundred-and-one stories of the future that I have read and can remember, excluding those by Wells, two stand out because they have great originality, depth, a touch of genius. The first—a book that deserves to be far more widely known than it appears to be—is *A Crystal Age* by that romantic naturalist W. H. Hudson. Most stories of the future merely carry our present trends to an extreme; they are huge exaggerations of what is happening now. But in *A Crystal Age* Hudson shows us a future that cannot be deduced from the contemporary world, a future as strange and fascinating, as alien to our conventional ideas, as the future might very well be.

The second of these books, better known than Hudson's, is *Last and First Men* by W. Olaf Stapledon, whom I remember with affection. For those readers who do not know this astonishing work, I must explain (though Stapledon begins his preface by saying "This is a work of fiction") that it purports to be a history of mankind going into an unimaginably distant future, 2,000,000,000 years hence. And it is an astounding feat of what its author himself calls "controlled imagination," full of wonders and marvels and yet not mere fantasy, in places strangely moving. (I read it first, over 30 years ago, sitting alone on a night railway journey, and now I suggest that sleepless night travelers by jet plane should try it.)

But although Stapledon introduces various "Time scales" into his text, his own interests were sociological and philosophical. Which brings me to the point I wished to make—namely, that in both these outstanding stories of the future, masterworks of their kind, the genuine Time element hardly exists. They are about Man, not about Time.

There is another reason why any account of fiction and drama about Time is difficult. It is possible that stories and plays may have been written by Time-haunted men in many languages. But from a literary point of view, they may be regarded as minor works, not sufficiently distinguished—or, on the other hand, not popular enough—to be translated for readers or playgoers in other countries. Some Polish or Yugoslav or Brazilian novelist or dramatist may have written something that would interest us here, because of its concern with Time; but to discover and include all such writers, then to examine and report on their work, would throw this chapter out of all reasonable proportion.

I cannot pretend, then, to be all-inclusive, surveying the whole world-field; all I can do is to look at various examples, those coming easily to mind, of Time fiction and drama, and to examine them in the light of our inquiry here. After all, Time is our subject—or, perhaps better, Man's relation to Time—and not fiction and drama.

It has been said that "modern literature" begins with Baudelaire, and in more than one place Baudelaire declares his fear and hatred of chronological time, the devouring dark enemy. A heightened consciousness of Time seems to arrive with this modernity. Henry James actually makes use of Time themes. From his earliest to his last poems, W. B. Yeats is throughout intensely conscious of Time. An American critic has analyzed T. S. Eliot's poetry in terms of three time levels. Thomas Mann, especially in *The Magic Mountain*, suggests the subjectivity of temporal flow, now running fast, now sluggish, according to our kind of experience, as his Hans Castorp discovers on the mountain. Joyce's Stephen Dedalus and Virginia Woolf's Orlando exist in more times than one. Among contemporary American novelists, both Thomas Wolfe and William Faulkner are clearly Time-haunted men. And so in England, I would say, is that strangely under-valued romancer, rhapsodical yet wise, John Cowper Powys. But the supreme Time novelist of our century—and a great novelist on any reckoning—is undoubtedly Marcel Proust.

Proust's vast novel is of course about people, for the most part belonging to a fashionable closely-knit society, but it is also about Time, as its title, *A la Recherche du Temps Perdu*, immediately suggests. In his youth he must have been strongly influenced by Bergson, to whom he was related, for it was while Marcel was

growing up that Bergson was more in fashion than any other French philosopher before or since his day. But it is both simpler and more rewarding to ignore Bergson and to consider Proust in terms of our inquiry. He is, I repeat, a great novelist on any reckoning, but he does owe much of his originality, appeal, and depth to the fact that he is really a novelist of multiple Time. It is a fact that he never admits; he never boldly withdraws from the accepted scientific-rationalist scheme of existence accepted by most contemporary intellectuals; and over and over again (but especially in his final volume) he is compelled to be ambiguous, clinging hard to metaphor in order to avoid making definite statements about Time.

It is possible, of course, that he had no such statements to make, that in theory he could not reject the accepted idea of Time, and this would explain the wistful, nostalgic, paradise-lost feeling that pervades his work. It would also explain why he has to make so much of "involuntary memory," not only bringing back forgotten scenes and persons but giving them a meaning, a

Above, the frontispiece to *Utopia* (1516) in which the English statesman Sir Thomas More describes an ideal state, with its communal property, religious toleration, etc. Today such "utopian" societies are often portrayed in stories of traveling in space or time. Right, an illustration from an early example of science fiction: *A Plunge into Space* (1891) by England's Robert Cromie—the story of a trip to Mars.

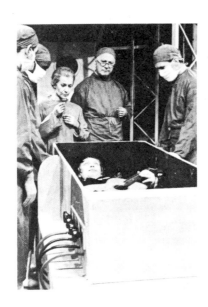

By transforming today's potential evils into tomorrow's realities, science fiction often contains a serious element of social criticism or satire. Left, in a scene from *The Bedbug* (1929) by Russia's Vladimir Mayakovski, doctors defreeze Prisypkin, who has survived in ice for 50 years. Below left, Prisypkin reacts with horror to the "perfect" communist state of the future.

Many of the stories by America's Ray Bradbury are also set in the future and underline the menace of technology and state control. For example, in the nightmare world of Bradbury's *Fahrenheit 451*, books are illegal and are banned and burned by the state. Below, in an illustration from the book, the fireman Montag unwillingly helps to burn down a library.

"Robots," which were first introduced to the world in 1920 by Karel Capek's play *R.U.R.*, have become a familiar feature of the futuristic worlds of modern science fiction, especially in stories that tend to "deify" the machine. Left, in a scene from Capek's play, the robots have triumphed over their human inventors, of whom only one has survived.

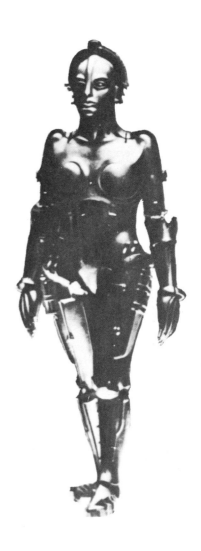

Left, the fabulous city of the future portrayed in the German film *Metropolis* (1925), partly a moralistic attack on the false glorification of technological power. Machines are represented as a source of evil and unhappiness —for example, the invention of a (female) robot, above, is partly responsible for rebellion among the city's serf-like workers.

symbolic depth, a strange beauty, they never had before; for he has to overwork this idea just because he will not openly accept any notion of multiple Time. But it is there in his vast wonderful novel, in which, we might say, life in Time One is retold from Time Two, with some hints here and there of Time Three.

From our point of view here, any attempt to describe life in terms of multiple Time is far more important than merely making use of a Time shift or twist to complicate an action, as for example John Buchan did in his ingenious *The Gap in the Curtain*. Actually there are few full-length novels with Time themes, generally concerned with characters moving from the present into the past or the future, but they have been fairly common in short stories. (There is, for instance, an excellent collection of such stories in *Travelers in Time*, edited by Philip Van Doren Stern, an American publication that I hope is still in print.)

Ouspensky, whose theory of recurrence will be examined in Chapter 4 of Part Three, tried his hand at fiction (in order to illustrate his theory dramatically) in the *Strange Life of Ivan Osokin*, in which a foolish young man begs an old magician to allow him to live the last 12 years of his life over again, so that he can avoid all the mistakes he has made. The magician agrees: "You will go back twelve years as you wish. And you will remember everything *as long as you do not wish to forget.* . . ." But, of course, the young man does wish to forget and so makes the same mistakes. The tale that follows is no masterpiece of fiction (and if he re-read it Ouspensky should have remembered his rather arrogant dismissal of writers in other places); but here and there, in its suggestion of multiple Time, it has some good moments.

A more subtle attempt to describe life in multiple Time—though as completely divorced from theory as Ouspensky's tale is wedded to it—is a short novel that I am glad to mention if only because it was published during the war and is probably little-known: *The Devil in Crystal* by Louis Marlow (Louis Umfreville Wilkinson). Without any explanation or theorizing, it describes how an easy-going intellectual and

sensualist suddenly discovers, on a morning in 1943, that he is back in 1922. He has then to re-live much of his life of 21 years earlier while still retaining his self of 1943. The relation between his outward existence and his most inward thoughts and feelings is suggested with uncommon skill and subtlety.

This is no major work for the critics to fasten on to; even though so short, it is episodic and rather sketchy between its brilliant opening and closing scenes; any attempt at a parable may have been far from its author's conscious intention (he probably regarded it as an unusual little *tour de force*). But any Time-haunted reader may find in it a sort of parable, if only because it may come close to the kind of experience we might possibly know in multiple Time—not now perhaps, but *then*.

Something must be said about science fiction, which for some years now has been so sharply attacked and so hotly defended (a sign at least that there is life in it). Following the example of its earliest and greatest master—the H. G. Wells who could write *The Time Machine* in his twenties —much of this branch of fiction has gone time-traveling and indeed has played all manner of Time tricks. But I am afraid that the results, from our point of view here, have been far more often bad than good.

It is true that science fiction can be an admirable medium for oblique satire and protest. We ought to welcome and applaud those writers who deliberately create the kind of world so many scientists and technologists believe they would like to live in, who by a bland exaggeration condemn implicitly some of the worst tendencies of our civilization. There are some good men at work in this field—notably Ray Bradbury, Simak, Kuttner, Asimov, in America; Arthur Clarke, John Wyndham, Brian Aldiss, in Britain. These and several more are not merely working with a little box of tricks: They are genuinely imaginative. It is not they who simply play cops-and-robbers in star systems with names that cannot be pronounced, defying credulity with some mysterious "overdrive" that makes the speed of light a mere crawl. Not for them conventional war stories with dubious scientific

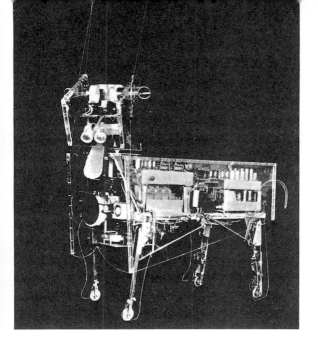

Above, the "Philidog," complete with electronic sensory organs and nerve systems—a real-life robot constructed in France in 1928. In much science fiction, the superiority of robots often seems to have made man redundant—an idea suggested in the Charles Addams cartoon (below) of endlessly self-reproducing robots: The scientist is saying to his companion: "Sometimes I ask myself, 'where will it ever end?'"

and technological trimmings and unbelievable galactic backgrounds. There is in the best of these science-fiction writers an element of satirical fantasy that mature readers should enjoy and not despise. Anybody who considers their finest work so much routine mass entertainment is a fool.

Nevertheless, after going so far with the defenders of science fiction, I must repeat my belief that the general effect of this form of writing has been mischievous. Much of it encourages some of our worst attitudes of mind. It takes to distant planets men who should not be taken to the nearest big town—aggressive and unbalanced 12-year-olds in possession of "atomic blast guns." (The bad science-fiction writer is the great *unconscious* satirist of our age.) It not only shows us robots but all too often shows us nothing else but robots. Man in these stories is more than halfway toward becoming a machine. Love and tenderness, art and philosophy, even humor and irony, all vanish from their future worlds, at which we stare in horror.

This may be all right (though possibly a trifle dangerous) when the writer has a conscious satirical intention; but very often he has not, and is simply describing life as he and his more devoted readers prefer to see it. So narrowness, one-sidedness, a foolish technological *hybris*, are confirmed and encouraged. The universe itself is robbed of mystery, wonder, and awe. The constellations and the depths of space are cheapened and vulgarized, as if idiotic advertisements had been scrawled across them. Planetary systems are turned into commercial depots and fun fairs. Our aggressiveness, predatory greed, and vulgarity are projected from here to Arcturus. It is our age at its worst.

So I for one deeply resent most of science fiction's monkeying with Time. (And here, I feel, some of its better writers are equally at fault.) It is bad because for many young minds it turns the mystery and wonder of Time as it really is into a storyteller's slick device, a mere "gimmick." The ancient enigma, masking a strange depth of experience, becomes an easy way of bustling cardboard characters through thousands of years, from one Hollywood-style civilization to another. (Not that I object to time traveling as such. After all, over 30 years ago, I wrote a short story, *Mr. Strenberry's Tale*, in which, in a tale within the tale, a man was trying to escape from the distant future into the present time, though he did not succeed—an important point.)

But time traveling should not be treated as if it were like some familiar railway line, packed with commuters. True, there must be some hocus-pocus, but a storyteller should claim only his modest ration. Wells, in *The Time Machine*, after a brilliant brief argument claiming Time as the fourth dimension, soon does some quiet cheating, not because he shows us a machine that can travel through time (this is fair enough, in fiction) but because he allows his time traveler to quit his machine and move about in a time in which he could not exist. Strictly speaking, this was cheating, though I do not suppose many of us object to it. But he did open a door, now wide open, to all manner of time-traveling nonsense.

We might as well take a look at this time travel, not to be pedantic about it but because it may apply to matters more serious than popular storytelling. John Robinson, we will say, was born in 1900. Now there may well be some part of his mind that is not entirely contained within chronological time. It might be possible therefore (and the claim has often been made) for Robinson to observe what is happening outside his own chronological time. If 1890 still exists somewhere, is still going on in some form or other, he might be able to pay it a visit.

But he could do this only as an *observer*, outside the scene and not acting in it, for 1890 plus a physical intervening John Robinson would not be 1890. If he wanted to do more than stare at 1890, if he wanted to experience 1890 as somebody alive then, he would have to make use of the non-temporal part of his mind to enter the mind of somebody living in 1890. (It may be that even as an observer he has to do this.) And of course some conscientious novelists—and really I am still thinking about writing—have realized this limitation, and have taken it into account in shaping their stories of time travel.

Remember, it is not the travel but the destination-time that enforces this limitation. It cannot be eluded by some S.F. hocus-pocus about reducing Robinson to his constituent molecules or atoms, blowing them along some time track,

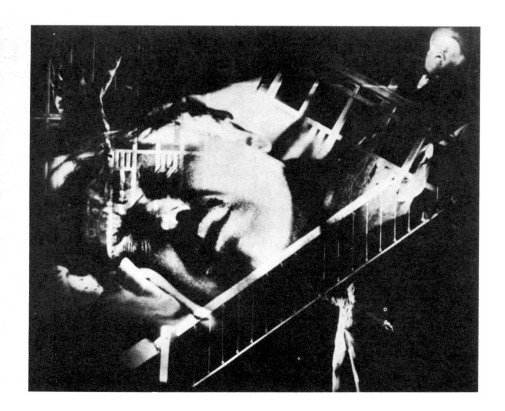

Left, a dream sequence from the German film *Secrets of the Soul* (1926). Superimposed images are used to suggest the dream's fantastic world—the sleeping mind's confusion of association, recollection, even precognition —in which past, present, and future mingle indiscriminately.

for Robinson to be reconstituted at the other end. The fact still remains that, except for the non-temporal part of him, Robinson does not belong to that time, whether it is 1890, 1490, or 2190, and as a physical entity, arriving whole or in parcels of molecules, cannot exist in it. If Robinson is born in 1900 and dies, let us say, in 1970, these 70 years of chronological time are all he has as one of the creatures of this earth, as a material being concerned with matter. And as such, he cannot be taken outside these 70 years, 1900–1970 being his limits in chronological time. (He may of course find himself existing in *another kind of time*, but we shall come to that in Part Three.) What I am really suggesting here is that while a modest amount of cheating is permissible in fiction (I have done it myself in several tales) a willful disregard of all possible limitations, now common in science fiction, is not good either for storytelling or our understanding of Time.

During the last 50 years a number of light whimsical plays, designed to please the average playgoer, have made use of some time shift, conjuring their central characters into the past or the future or into some might-have-been situation. Their authors, I imagine, would have been the first to admit they were not concerned with the Time problem. To search for dramatists who are seriously concerned with it is very difficult and unrewarding. Playgoers who find the Time problem fascinating must not be surprised by this scarcity. Any serious treatment of a Time theme in the theatre has to discover a way past some alarming obstacles, unknown to the writer of fiction. (Here the film would be by far the better medium, that is, in itself, without any regard to the cost of production or the prejudices of distributors.)

For example, a highly subjective theme, like that in *The Devil in Crystal*, is almost unworkable on the stage. Something could be done with trick lighting and with recorded voices speaking the interior monologues, but the result would probably be clumsy or tedious or both. There is something obstinately objective about stage production; the actors in it have solid bodies and need time to change their clothes and make-up; and this explains why the film, with its dissolving images, is so much the better medium.

There is another and more important reason why the serious Time play is so difficult. A

The confusion of Time in dreams has often been put to good use in fiction: Right, an illustration to a story from Boccaccio's *Decameron* (first published in 1353) depicts the rescue of a woman attacked by a savage wolf. In the story, the event had been foreseen by the woman's husband in a dream.

novelist offers his work to a number of individual readers, who attend to him at their own convenience, reading quickly or slowly as it suits them. If there is something they do not understand, they can read the passage again and try harder. The dramatist, however, has to deal with an audience, perhaps a thousand people of all kinds, some very clever, some very stupid, whose collective attention and sympathy he must win and keep. If he explains too much, he will bore the clever playgoers; and if he cuts his explanation short, to please them, then he will bewilder and finally bore the stupid playgoers. He might possibly find a Time theme that explains itself, but the odds are heavily against his doing anything of the sort.

(It is worth pointing out that unlike his predecessors, a Shakespeare or a Calderon, a Racine or a Schiller, a serious modern dramatist cannot assume that he and his audience share any common ground of belief and opinion, any accepted code of morals and manners, any *Weltanschauung*—at least in the West. In the communist East, he suffers from the opposite misfortune, he and his audiences being too closely packed into their common ground.)

If the idea of multiple Time came to be widely accepted, then plays on Time themes (even though the production difficulties remained) would be far easier to construct. As it is, the task of conveying the central idea, so that the drama can be understood, remains a baffling obstacle. And this, I am certain, is why Time plays, except on a level of light entertainment where the time shifts are not taken seriously, are so very hard to find. They exist in manuscript—as I knew only too well in the later 1930s—but not where plays belong, in the theatre.

I have never seen a production of Henri Lenormand's *Time is a Dream*, but having read it with an experienced eye, I can see that it could be theatrically effective. Here, as in his other plays, Lenormand is the dramatist of a corrupt atmosphere and various stealthy and sinister

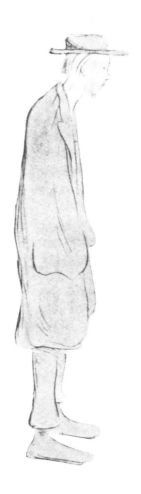

Many stories revolve around a longing to recapture the past or to foresee the future. Left, a drawing of Enoch Soames, a third-rate poet of the 1890s who (in a story by Max Beerbohm) sells his soul to the devil for a visit to the British Museum in 1999. Above, two "Time-jumping" 19th-century lantern slides depict (left) a man's reverie of lost youth and (right) a young girl's vision of future romance.

Right, a scene from *Dangerous Corner*, J. B. Priestley's first play, which concerns a "loop" or "split" in Time. The play opens after a dinner party where a casual remark by one of the guests precipitates a series of unpleasant disclosures about the characters. The play then "begins again" with an alternative version of the same after-dinner conversation—which this time is undisturbed by any dramatic revelations.

devices. While walking near a lake, a girl has a waking prevision (they are not uncommon) of the death by drowning of the man she is going to marry, now returning from the East Indies. She does not tell this man, Nico, but we soon discover he is a pessimistic fatalist, who tells her—"What every one of us must learn to forget. Our destiny, our fate, the idea that although we might at a stretch believe ourselves free in space, we know we are the prisoners of time. In Java you do learn after a while to put all that out of your mind." But in Holland, where the action takes place, he seems to unlearn these Javanese lessons very quickly, barely concealing his longing for death, and declaring that to die is to awaken from the dream of time.

By one artful stroke after another, Lenormand brings the actualization of the tragic prevision nearer and nearer, until we know at the final curtain that now Nico must be drowning in that lake. All this can be theatrically effective, but *Time is a Dream* does not seem to me a satisfactory Time play. I doubt if Lenormand approached the writing of it through any consideration of Time. As with earlier and later plays (this one arrived in 1919, about the middle of his career) he was fascinated by the idea of his characters being helpless puppets of fate. Nico and his girl and their friends are carried, as if by a conveyor belt, to the tragic point in the fixed future. In Time terms, this is too easy and obvious a dramatic device, and anyhow is wrong. Even if Time should be a dream, it is a much more complicated dream than that.

Thornton Wilder, more than most American dramatists, is haunted by the idea of passing time. Perhaps his best expression of it is in a one-act play, *The Long Christmas Dinner*, in which the lives of several generations are compressed into an hour or so at the Christmas dinner table, with the old retiring from it and the young arriving—an unpretentious and touching little play, more effective, in my opinion, than his longer and more ambitious pieces.

Among the might-have-been plays, not unrelated to the Time theme, the best I know is Barrie's *Dear Brutus*. Because Barrie was over-indulgent in his sentimental moments, we tend to forget not only what I have called elsewhere his "goblin cunning in stage-craft" but also his sardonic humor, the frequent bleak cynicism that was the reverse of his occasional sentimentality. Though one character in *Dear Brutus* may have been rescued by having a second chance, the others are no better off or even worse. Barrie was not always writing *Peter Pan*, as too many people seem to imagine, and a close study of his plays (including that sardonic little masterpiece *The Will*) would reveal many strange elements in them—one of them, I fancy, involving Time.

Now a personal note. It has always seemed to me bad manners, an insensitive display of egoism, for editors of anthologies, writers attempting critical surveys, and others, to include or make any references to their own work. But here (as my publishers insist) we have a special case, with subject not merit the first consideration. So, feeling a trifle awkward, I will end this chapter by pointing out that in *Dangerous Corner*, a box of tricks that has had more success than it has deserved, I play with the idea of a loop or split in Time; that in *Time and the Conways* I make some use of Dunne's theory (which I will discuss in some detail in Chapter 10, Part Three), and in *I have Been Here Before* I press into service Ouspensky's theory of recurrence (see Chapter 11, Part Three); and that—for they might as well be brought in for good measure—in two experimental plays, *Johnson Over Jordan* and *Music at Night*, I attempt a kind of four-dimensional dramatic technique, with scenes outside passing time, like those we know in our dreams. As for fiction, I have been less ambitious there, but the Time theme does make its appearance in a novel called *The Magicians* and in some of the short stories collected under the title *The Other Place*. And now I *with a blush, retire*, at least from this Part One, which anyhow has come to an end.

In *I Have Been Here Before* (1937), Priestley bases the action on Ouspensky's theory of Time and human life as a recurring cycle. Right, Walter Ormund, one of the characters who is saved from an otherwise tragic course of action by the mysterious Dr. Götler (seen above explaining his precognitions).

Below, the Conway family in Act I
of Priestley's play *Time and the
Conways* (1937). Above, the same
characters in Act II after 19
years of disillusion and failure.
The third and final act repeats the
happy family party of Act I—a
"Time twist" that sheds an ironical
light on the youthful aspirations
of the characters, particularly
those of Kay Conway (right).

Part Two   The Ideas of Time

5   Time and Ancient Man

An aboriginal rock painting from northwestern Australia. The central figure—"the emu of eternity"—is one of the mythical figures whose existence was wholly outside ordinary passing time. Such figures dwelt in the eternal, sacred "Great Time" of myth, ritual, and dream.

5 Time and Ancient Man

1. Primitive Man

Our knowledge of the past, especially the distant past, has increased enormously during the last few decades. Moreover, since the end of the Second War, public interest in this knowledge has increased enormously too, as many publishers' and booksellers' catalogues can testify. There is a wide market for large and expensive books on ancient history and prehistory that did not exist 25 years ago. It is as if Western Man, less sure of himself than he used to be, were becoming uneasily curious about vanished civilizations and those remoter ancestors of ours who took time off from their roaming and hunting to bequeath us their cave paintings. There remains, however, even among some of the researchers themselves, a good deal of what might be called technological bias and snobbery, against which we must all be on our guard.

We are at the moment hypnotized by our mechanical inventiveness and technological skill, so spellbound that we no longer demand that they should serve us—we have to serve them. (If an example is needed, then consider supersonic aircraft, which we do not really need for civil transport, which will probably be dangerous and certainly a horrible nuisance, but which the *Zeitgeist* will soon compel us to have.) We can afford to take some pride in our mechanical inventiveness and technological skill, but we cannot afford, not at this dubious moment in our history, not with so many frustrated and anxious people all round us, to make this inventiveness, this particular skill, the test of human accomplishment. To put it another way: An ancient people—the Incas, for example—might never have arrived at inventing the wheel and yet might possess some virtue, perhaps some particular grace in the art of living, that we do not possess.

It is true that we in this age have one advantage over all older civilizations—namely, that a far wider and longer vista is open to us; we really know far more than they did. Nevertheless—and this is the point I am anxious to make—this greater knowledge may be useless, may even be dangerous, unless we are prepared to make an imaginative effort.

The point had to be made just because we have now to consider other peoples' ideas of Time. An imaginative effort is particularly necessary here because we are now fixed in our own idea of Time. Our minds are so clamped to the notion of irresistible *passing time*—and I use that term with grateful acknowledgment to the late Maurice Nicoll, whose *Living Time* I hasten to recommend—that we shall have to wrench our minds away to understand what other peoples thought and felt. This effort is particularly necessary when we are considering primitive man.

And here, I feel, another point must be made. We have to begin by appreciating, and so making a large allowance for, the *undifferentiated* state of the primitive mind. This is not always understood by the very men who ought to understand it. For example, what about the famous cave paintings at Lascaux? Are they religious, are they art, are they magical devices to bring success to the hunt? Merely to ask such questions is wrongly to endow primitive man with our own intellectual powers of analysis. These paintings were not this, that, or the other, but an expression of an undifferentiated totality, not to be divided into religion, art, magic, and economics until after the lapse of many thousands of years. In the same way it is useless to bring to primitive ideas of Time our own kind of intellectual analysis, pointless to inquire "what exactly did they mean?" In our terms, they didn't know; so we can't know.

Except at certain heightened moments, we feel that whatever we may be doing we are contained within passing time. The vast invisible clock ticks on and on. Primitive man had no such feeling. It was not passing time but the Great Time that had meaning and reality for him.

Right, a 17th-century engraving of Shakespeare's Sir John Falstaff, one of the great characters of literature. Such figures seem to have acquired a separate existence, outside literary works: They are, we might say, the mythological beings of Western man (and thus exist immortally—without aging—in our version of the eternal Great Time). Top right, friends of Charles Dickens mourn the death of Little Nell (in *The Old Curiosity Shop*) as if she had really existed. Center right, a child's letter addressed to Maigret, the police inspector created by Georges Simenon, and the hero of a British T.V. series.

Below, a carved wooden "churinga"—a dwelling place of the Australian aborigines' ancestral spirits. The aborigines regard their sacred objects and rituals as links between ordinary life and legendary dream time. Below right, in an aboriginal ritual a tribesman sings a sacred song.

MR MAIGRET
TV companies
BBC London

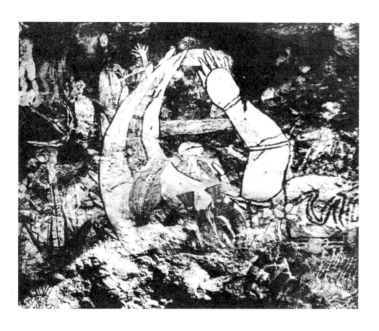

Two examples of aboriginal rock paintings (from Arnhem Land) provide analogies to the aborigines' concept of the Great Time as a whole in which past, present, and future co-exist. Above, a kangaroo's tail and claw (recently drawn) superimposed on a background of more ancient work; below, new drawings of European ships mingle with older traditional designs.

The Great Time belonged to the myths, the gods, the mighty spirits, the magical creators; and when primitive man worshiped, performed ceremonial rites, or engaged himself in any meaningful act—for when he went hunting or fishing he might be deliberately imitating some mythological hero—then he entered the Great Time. His world was sustained not by the meaningless passage from dawn to noon and noon to sunset, but by this Great Time, in which all creation, power, magic, existed.

Probably the best remaining examples of primitive men are the Australian aborigines, whose beliefs have been scrupulously studied, notably by A. P. Elkin, to whom I owe much of what follows. (All the quoted passages are taken from Elkin's *The Australian Aborigines*.) Now these people believe that every man possesses a soul or spirit that exists eternally. The life of these spirits is in cyclic form: They "pre-existed usually in definite sites in the country of the group"; then, incarnated through the mother, they enter profane life, passing time. At death they return to their spirit-homes.

And these dwelling places exist not only in a sacred world but also in a sacred time (the Great Time)—a time that is qualitatively quite different from profane or passing time. It is "*all-at-once* instead of *one-thing-after-another*," past and present and future merging and becoming one—the eternal instant.

Their myths relate the deeds of great ancestors or heroes. "To the aborigines, things are as they are, because of the personal actions of heroic beings in the past." But this "past" belongs to that same sacred time of the spirit homes: It is also present. The usual term that they use for the past creative period also means "dreaming": Their myths are about the eternal dream time. "The time to which they [the myths] refer partakes of the nature of dreaming; as in the case of the latter, past, present and future are, in a sense, coexistent—they are aspects of the one reality." They have many forms of totemism, embracing different kinds of life, but their term for the cult totem is also the "dreaming"; and the cult totem is the "door into the eternal dream time." The opening of that door is through ritual, its ceremonies abolishing profane or passing time and leading the way into the eternal dream time. When in these ritual ceremonies the original great deeds of mythical heroes are re-enacted, the men taking part in them enter the time when the deeds were originally performed, not into the past but into an eternal present.

Access to the "eternal dream time," these aborigines believe, can be gained only by observing the traditional rules of behavior. The "secret life" of ritual and mythology is infinitely more important and meaningful than day-to-day existence, earning a living, trying to appear important, and so on.

"It is the life," as Elkin tells us, "in which man really finds his place in society and in nature, and in which he is brought in touch with the invisible things of the past, present and future." From it men obtain courage and strength, but not if they "desecrate or neglect the sites, break the succession of initiates, forget the myths and omit the rites," all of which results in the loss of "an anchor in the past, a source of strength, and a sense of direction for the future."

When a young man is initiated into the sacred life of the group or tribe, he is given a new name, usually taken from one of the myths, and this name is never mentioned except on sacred ground. It is his passport "into the eternal, unseen world of ancestral and totemic heroes." And now the initiate plays his part in keeping the group or tribe aware of that other world, that other time—the eternal dream time.

So much for what is—or has been—believed by black men coming to us out of the Stone Age. (Somebody should compare these beliefs, in terms of breadth, depth, and response to the profoundest psychological needs of man, with the beliefs of the average citizens of Sydney and Brisbane.) Here the indifference to passing time and the emphasis and dependence upon the Great Time are obvious. This is modern man's life in reverse. He feels his life is being ticked away—this is the only reality—and cannot enter into any "eternal dream time," knows nothing of the "invisible things of the past, present and future," and has no such hidden source of courage and strength: He feels himself fastened to a hawser that is pulling him inexorably toward the silence and darkness of the grave.

Yet he too dreams, discovering, if he chooses to remember it, in this strange night existence a confusion of past, present, and future, as we shall see. But no idea of an "eternal dream time," where gods and heroes (from whom he is not separated for ever) have their being, comes shining through to make modern man forget his calendars and clocks, the sands of his time running out. Outwardly the black fellow, scrabbling in the desert for edible roots and small animals to cook, cuts a poor figure compared with the white man descending from his aircraft to ride in a car toward a Melbourne expense-account lunch; but inwardly, where the psyche also needs sustenance, the case is altered, and it might be better to drop the comparison.

We have come a long, long way from those ancestors of ours whose ideas of Time cannot have been very different from those of the Australian aborigines. Passing time, once almost meaningless, is now the inescapable beat, like that of the engine of some space ship, of the whole vast universe; we seem to be utterly at its mercy; while any idea, once so all-important, of the Great Time, the eternal dream time, the other time of gods and heroes and mythology, seems to have vanished.

But in fact it has not vanished entirely. How could it? After all, we are the descendants of these primitive men, linked to them through blood and bone and nervous systems and various mysterious areas of the brain, and, however much we may despise it, we cannot escape our inheritance, must accept what our ancestors have bequeathed to us. And because there were countless generations of them, long preceding for hundreds of thousands of years the very earliest civilizations, what we have inherited from them is, so to speak, buried deep into us, beyond the reach of our skeptical rationality. In spite of our conscious rejection of the distant past and our towering *hybris* (indeed, perhaps because of them), we are haunted men.

Among the ideas that haunt us—ideas we may laugh at but that will not leave us, ideas that often promise a mysterious happiness when all else seems to fail us—is this one of the Great Time, the mythological dream time, that is behind and above and altogether qualitatively different from ordinary time. We no longer create any grand central system out of it. We do not let it shape and guide our lives. It has dwindled and now looks small and shabby, rather laughable; but it cannot be laughed out of existence, it refuses to go away.

So, perhaps with a shrug, smiling at our folly, we turn to this idea of the Great Time—in the hope of some relaxation—when what has happened in passing time (really all we have, we feel) has left us battered and shaken. It is in fact in the fiction and drama offered as popular entertainment and so much relaxation that we encounter these ancient haunting ideas.

So our best-known detective story heroes and their assistants do not exist with us in passing time; Erle Stanley Gardner's Perry Mason, Della Street, Paul Drake, for example, have remained the same age for over 30 years; such characters, belonging to their own kind of mythology, live like the gods and heroes of primitive man in the Great Time. Some popular storytellers, determined to defy this ancient tradition and to be realistic, have allowed their heroes to remain in passing time and grow older with their readers; but they have regretted it. So Agatha Christie, aging and then retiring her detective Hercule Poirot, soon had to conjure him out of our time into the Great Time, where he remains, oldish but never any older.

The idea of the eternal dream time haunts us in curious ways, moving the minds and actions of men who would sharply deny they ever entertained any such foolish notions. We can take two examples from the glaring giants of the West and the East. There exists in America, especially among commercially successful but hard-pressed men, a myth of the return-at-last-to-the-little-home-town, to put on old clothes and go

fishing again as you did when you were a kid out there. And here, barely disguised, with just sufficient covering of reality to escape derision, is the longing to get out of passing time into the eternal dream time, where there is no Wall Street, no Madison Avenue, no ulcers or high blood pressure, where little home towns remain unchanged and the boy-men fish for ever.

Behind the Iron Curtain, although the demands of technology are much the same so that men are equally more and more grimly aware of passing time, the ancient haunting idea takes a different form. Many an apparently dour Russian communist in his secret heart looks forward not simply to a society materially better than the one he knows but to an immediate future, just around the corner (where it will always be), altogether different from anything man has ever known. And what he is hoping, though he would not (probably could not) ever admit it, is that by some magic, perhaps born of a vast communal wish, passing time will be transformed into the eternal dream time.

But for all his initiation ceremonies, odd rites, sacred objects, superstition, and ignorance, primitive man knew better, believing as he did that in ordinary meaningless time a man could not return to his past, as the American dreams, or look forward to a magical future, as the Russian does; but that the Great Time, the eternal dream time, meaningful and ever-enduring, is here with us, if we do not deliberately blind ourselves to it, and is ready to give us courage and strength.

A ninth-century Javanese carving of a Hindu god of Time. The god is known as the "Black One" and, like the Greek god Kronos (identified with Time) who devoured his young, was seen as a relentless destroyer of the universe.

2. Mesopotamia and Egypt

When we come to the earliest civilizations, represented here by Mesopotamia and Egypt, it seems to me that a greater imaginative effort is demanded of us than when we are trying to understand primitive man. Now we have left isolated groups or tribes of hunters and are considering large ordered societies, elaborately organized, able to build and administer cities and, as in Egypt, to erect vast structures, still after thousands of years among the wonders of the world. So we shall find it hard to believe that the Egyptian pyramid builders or the Mesopotamian astronomers were not far closer to us in their modes of thought than they were to primitive men.

They knew, as we have seen, how to measure time. Therefore we feel that their ideas of Time cannot have been very different from ours; and when we are told that in fact their ideas were altogether different, we begin to wonder, not without some feeling of irritation, how much we must allow for mere fancifulness and *pretending*. And then we shall go badly wrong.

To this we shall return, but I must insert here a private query and doubt concerning ancient Egypt. For 3000 years it endured, experiencing some upheavals but on the whole enjoying a tranquility, a settled course of life, a freedom from anxiety, that no other civilization can show us. In spite of its extraordinary concern with death and the possibility of life beyond the grave, and a certain official gloominess, its folk appear to have been hard-working, cheerful, lively people, fond of beer and pastimes. But high and secure above them was a privileged priesthood, with more opportunities down all these untroubled centuries for contemplation and speculative thought than any other men in this world can ever have enjoyed.

Moreover, they were under no obligation to share their discoveries with anybody outside their order. Now I know that at least 50 per cent of our crackpots and mystery-mongers, ready to find the Great Pyramid bristling with prophecies, have decided for ancient Egypt (perhaps in its capacity as a colony of lost Atlantis) as the home of profound esoteric knowledge, a secret wisdom. Yet we cannot be sure such hidden knowledge did not exist.

The Egyptian priests had exceptional, perhaps unique, opportunities to acquire and pass on such knowledge. They also had a formidable reputation as magicians, seers, powerful wise men, throughout the ancient world—a reputation often recognized and acknowledged by the Greeks, who were no fools. It is quite possible then that far more was known in ancient Egypt than has been revealed to us.

What is certain, from all that has been revealed, is that in spite of their social organization, their elaborate religious beliefs, their cities and temples, these men of the earliest civilizations were far closer to primitive man in their modes of thought than they are to ourselves. They were still living in a realm of myth.

This does not mean they made up stories and gave fancy names to what we now recognize as natural laws and processes. To them there were no such laws and processes; there was no machinery, so to speak, entirely separated from man. Instead, there was an endless drama of personal or divine will and action.

Nothing happened mechanically; somebody was doing something to you. The sun was not obeying a law; it was a god beneficiently appearing, working a miracle every morning. One day, if feeling not sufficiently appreciated, he might not appear. If there was a storm, someone very powerful was angry. Everything on earth or in the sky was alive: There was no "it." And in such a world the myth, intensely personal and pictorial and concrete, was a way of thinking, in fact the only way.

Too often this kind of world and what is called its "mythopoeic" thought are patiently explained to us as if the whole subject were very difficult, being all so remote from us. Now this may be true for the scientist or the drier kinds of scholars. But the fact is that this ancient world and its way of thought still exist in us, beneath the level of what we have acquired from our own age. Poets, imaginative men, most women, almost all children, know them well. We have only to be shaken by strong feelings, of happiness, misery, terror, to imagine again that all the

"it" we have taken so complacently has changed character and quality, is now alive, personal, blessing or threatening us. (Great clowns like Grock or W. C. Fields got many of their best effects from apparently struggling with objects that had a malevolent life of their own. I doubt if we should laugh if we felt these situations were either true or utterly false; it is the escape from the vague feeling that they might be true that makes us laugh.)

Even with Time, on a certain deeper and more instinctive level, we abandon our idea of a universal process (the most impersonal process that we can imagine) and begin to think with resentment or gratitude about "bad times" and "good times"—almost as if some god had delivered them to us.

Nevertheless, though their world is still alive in us and we know something of their "mythopoeic" thinking every time we dream, we have to make an imaginative effort to understand these earliest civilizations. We have to rid ourselves of two ideas that have dominated our own age. For whatever else we feel, we feel that we are in History and we are in Time. And these ancient people in Mesopotamia and Egypt, for all their impressive achievements, were neither in History nor in Time.

Each people was intensely aware of a Beginning—first, the original chaos "before the sky was, before the earth was, before men were, before the gods were born, before death was," followed by the creation myth; but this Beginning was not regarded, as we might regard it, as a kind of switching on of Time and History. In a sense it was not even in the remote past. It was something that had to be constantly renewed through rites and ritual drama, in which rulers, priests, and priestesses did not merely play the parts of gods and goddesses but were thought to be temporarily identified with them.

So in Babylonia, which justly felt itself constantly menaced as Egypt never did, at each new year the original victory of the gods over watery chaos was solemnly re-enacted. There was no feeling here of the steady and inevitable movement of History. Life was not historical but dramatic and magical.

The ancient Egyptian "god of millions of years." To indicate his great age, the god is depicted as old and fat; he also holds a notched stick representing the countless years of eternity. His left hand is stretched out to protect an eye that symbolizes the sun on its nightly passage through the waters of the underworld.

As for their idea of Time, Henri Frankfurt has stated it admirably: "The concept of time as it is used in our mathematics and physics is as unknown to early man as that which forms the framework of our history. Early man does not abstract a concept of time from the experience of time." It is the absence of this concept of time, which now might almost be said to be built into us, that marks the difference between these civilizations and ours. Here, as in their essentially "mythopoeic" thinking, they are close to primitive man and far removed from us.

Nevertheless, with their settled agriculture and their ordered societies, they were very different from groups or tribes of hunters and could not be satisfied with the idea of a meaningless passing time and a meaningful Great Time. Their very pursuits and their social organization, with all its regular festivals and ceremonies at fixed dates, brought them closer to passing time. Men cannot live in cities and play their parts in some elaborate hierarchy, drilling armies or making estimates of the barley for the royal granaries, and take no count of time that does not open into the eternal dream time.

Yet this did not involve (unless it existed in esoteric knowledge) an approach to our abstracted concept of time—our empty and impersonal time. What they recognized was what Cassirer has called "biological time"—that is, a sequence of different phases, forever renewed (but by will, not automatically), both of man's

life, with its profoundly different "times" between childhood and old age, and that of nature, with the recurring rhythms of the seasons. (And man and nature had not yet been sundered; nothing was yet entirely objective, an "it.") These filled and varied "times" are a long way from our abstracted concept. Yet here again, in the warm depths of our personal life, below the level of analytical thought, we are still sharply aware of such "times," especially at moments when our concept of empty and impersonal time, the time that everybody has and nobody has, seems meaningless to us.

Toward the end of the Old Kingdom in Egypt, we are told, there were three time gods: one who brought storms, sickness, sudden death; one who gave life; and the third, uniting the opposites, representing the soul of the godhead.

They who circle daily on his [the sun god's] *path*
As living souls of the gods
They who serve the sun god
As messengers in the cities and countries.

Although it was said of these time gods that "Heaven, earth and the underworld are subject to their plans"—in which, it must be admitted, there is at least a hint of our later conception of time—they remained lesser gods, not responsible for some all-embracing and featureless Time but for various changing "times" that expressed their functions and characters.

And we are still not in History, set in motion somewhere in an ever-receding past. The task of each Pharaoh or Mesopotamian king, on his accession, was to renew, to restore in its triumph, the Beginning.

And he and his subjects were not in a smoothly running chronicle, but taking their parts in a huge drama—which might be the war of the gods against chaos or the yearly marriage of the goddess of fertility with the shepherd-god Tammuz—a drama that, the Mesopotamians knew, certainly might suddenly end in universal tragedy, the whole scene of human life suddenly vanishing into darkness and oblivion. Our grandfathers would probably have smiled at such childish fears of the immediate future. Now we smile no longer.

The people of Mesopotamia, at the mercy of unpredictable floods or droughts, felt themselves —and indeed all nature—to be at the mercy of forces outside man's life, beyond his control however devout his supplication. In spite of all their ceremonies, rites, and ritual drama, they believed that darkness and death (to them the supreme punishment) could still, at any time, win a universal victory.

The Egyptians, nourished so regularly and bountifully by their magic river, which to this day seems to have a character and atmosphere all its own, could afford a less stern or fearful outlook. Not only did they accord, as Jacobsen says, "to man and to man's tangible achieve-

In ancient Mesopotamia (as in ancient Egypt) the sun's daily progress was seen as an event in a recurring process dependent on divine will. Above right, a Babylonian seal (*c.* 2400 B.C.) depicts the sun's setting as the disappearance of the sun god Shamesh (center) watched by various attendant deities.

Right, the falcon-headed sun god Ra (crowned by a sun disk) with the goddess Hathor, from a 13th-century B.C. royal tomb painting. In ancient Egypt, the sun's course was seen as Ra's boat sailing across the sky. Below, in a drawing of the same period, a similar sun disk between two lions that symbolize today and tomorrow.

ments more basic significance than most civilizations have been willing to do," but while publicly emphasizing the fact of death—in sharp contrast with our own civilization, where for instance in America the very word can almost be said to be banned—such elaborate arrangements were made for life after death that this grimmest of realities lost much of its terror. At least this was true of important personages: Peasants had to take their chances in the next world as in this one. As Breasted remarks: "Among no people ancient or modern has the idea of a life beyond the grave held so prominent a place as among the ancient Egyptians." Life, certainly for the more fortunate, had always been so pleasant along the banks of the Nile, with its clear sweet dawns, its sleepy afternoons, huge burning sunsets and mild nights of stars, that surely it ought to go on forever?

This brings me to an idea, familiar enough to us, but surprising—and, I will confess, to me disappointing—when encountered here. We find it in a quotation by Budge from the ancient Egyptian *Book of the Dead*. There the deceased asks the god Thoth: "How long shall I have to live?" And Thoth replies: "It is decreed that thou shalt live for millions of millions of years, a life of millions of years."

In one jump we seem to have arrived in one of the shabbier corners of our own age. No esoteric

wisdom here. This is the so-called eternity of our Bible Fundamentalists, spiritualists, and the like —the false eternity that is simply our passing time running on and on and on. It is time as we think we know it without any qualititative change, only more and more and more of it. Though in this life beyond the grave we may be invisible to mortal eyes, existing somewhere behind a veil, we are still in Time and History; there has been no fundamental change. We are asked to live in a brutally simplified universe.

It may be that powerful and rich men, then as now so often fearful, vain, and egoistic, had to be cosseted even in the *Book of the Dead*. But certainly primitive men, like the Australian aborigines, could do better than this. The idea of the Great Time, the eternal dream time, with its qualitative difference, its *all-at-once* instead of *one-thing-after-another*, is infinitely more profound. Some wisdom seems to have been lost when the hunters and cavemen settled down to till the soil and then became not only farmers but civil servants, generals, high priests, soothsayers, and pharaohs. And a bad idea was born—the idea that eternity is simply a vast helping of passing time—an idea that makes mischief to this day. How many sensitive minds must have been first bewildered and then repelled by this false eternity offered them by evangelical Christianity, even the Heaven that would pall after so much repetition, let alone the nightmare of ever-enduring Hell, one of the most evil notions that ever entered the human mind?

I agree that it could be argued—and now a host of saffron-robed priests seem to look at me reproachfully over their begging bowls—that what the god Thoth was offering his questioner was not the false eternity of popular Christianity but the innumerable incarnations, our fate while we are still bound to the wheel, accepted by the Buddhists. This Egyptian, it could be said, would live for millions of years because he would return again and again and again to Time, in one shape and personality after another, until finally purged of all desire for any further existence on this earth. (On this theory, we may have met him, the head of a corporation or selling toys in the street.)

But even reincarnation (an idea with its own subtleties and not one to be rejected off-handedly) over-simplifies Time just as it does personality. Thoth's man was really asking how long he would live as himself in his own time, the time that had made him what he was, that helped to create for him that four-dimensional "long body" to which I referred earlier. And Thoth, who ought to have known better but who may have been misreported, as are so many celebrities, wildly simplified everything by multiplying by millions.

Indeed, the Egyptians can do better than this. I recall a hot afternoon when I was toiling and sweating up the Valley of the Kings. The dragoman, who was even fatter, hotter, and sweatier than I was, when he was not cursing and driving away the children who were asking for *baksheesh*, was explaining why the ancient kings had gone to such lengths to ensure that their tombs would be sealed and hidden. It was not, he said, because tomb robbery was at the time an immediate possibility, for the tombs could then be safely guarded. No, these elaborate precautions were taken because the soothsayers and magicians had declared that sometime in the future, perhaps after centuries had passed, tomb robbers would arrive to try to discover and to loot these sacred places.

They were not, I gathered, making a rational anticipation, saying in effect that sooner or later, when the kingdom was in disorder, there would be bound to be men who would want to loot the tombs. They were directly prophesying, after peeping into the future, that tomb robbers would arrive. Therefore, because the future—or some possible future—had been seen, then action had to be taken in the present in the hope of changing the future—although it was a future that in some form already existed.

Now I do not care if no such situation can actually be found in the history of ancient Egypt, or if instead this was just an idea in the mind of the dragoman, a rum character; it offers me more to think about than the god Thoth's millions and millions of lives. It comes closer, I believe, to the actual complexity of things. So Egypt redeems itself.

3. Greek and Roman

I have postponed until Chapter 6 any consideration of the ancient Indian vision of kingdoms and empires, worlds, whole universes, glowing and fading, then vanishing like pricked bubbles. This vision became, in more sober Greek thought, Time moving for ever in an ordered procession around the Eternal. But it is worth noticing that, especially among the earlier Greeks, an idea remained from this other stupendous cosmology.

The smallest complete cycle there, for India, was the *mahayuga*, consisting of four ages, each shorter than the one before it and each revealing less of the divine *dharma*. At the beginning of the cycle there was the *Krita yuga*, the perfect age in which men lived in harmony with the divine order of the universe, and at the end of it, the fourth and evil age, the *Kali yuga*. (We are in a *Kali yuga* now, coming indeed toward the end of it, teetering toward chaos and darkness.) But it was the idea of something like the first, longest, happiest of these periods, the *Krita yuga*, that lingered on in Greek thought and became the Golden Age.

This idea of an earlier time, an earthly paradise in which everything had been simple and good, has been one of our most persistent haunters. (We can find it even in Laotse's sardonic outline of history, which may be paraphrased: "After Tao is lost, we have kindness; after kindness is lost, we have justice; after justice is lost, we have social order and control, the thinning out of loyalty and honesty of heart and the beginning of chaos.") It is behind all the Arcadian shepherds and shepherdesses, the Strephons and Phyllises, of literature. It is there in Rousseau's discovery of the Noble Savage. Young enthusiasts still set out for the South Seas looking for it. One of our most dauntless contemporary theorists, H. S. Bellamy, tells us that between the final disintegration of the old moon (it somehow supplied the *Book of Revelations* with its apocalyptic images) and the earth's capture of a new satellite, our present moon, there was a considerable moonless interval, without tides and sinister bright nights—a quiet but happy time that man remembers as the Golden Age.

It is only by making a very determined effort that some of our technologists, both in the East and West, have torn the Golden Age out of the past and planted it in the immediate future, when 6,000,000,000 of us will live happily in concrete tenements in cities 100 miles long, clothed in plastics, nourished by gobbets of synthetic protein, and watched by, not watching, television. Even as a child I could never quite believe in that Arcadian sheep-raising, so clean, neat, and dry; but I prefer the past's Golden Age to that of the future.

We seem in many ways so close to the Greeks that we may be tempted to dismiss with a shrug their different attitude toward Time. The fact that they imagined it to be moving in a great circle may seem at first of little importance. But there is a profound difference between the cyclic and rectilinear ideas of Time. The Greeks did not see themselves in History as we do. True, they might disagree, as indeed they did, about the quality of man's life along the cycle. Some Golden Agers, like Hesiod, might believe that the world, new and shining, was never as good again as it was at the beginning of the cycle. Others (it may have been a matter of temperament) might reject this idea of a gradual deterioration and assert a steady improvement, like Xenophanes: "The gods did not reveal to men all things in the beginning, but in the course of time, by searching, they find out better."

This may look suspiciously like a sense of History. But it was not, in our understanding of the term. Nothing was happening for the first and last time. They were not on a road but a vast roundabout, with the eternal gods and their unchanging stars where the engine and the music are, in the center, unmoving.

They might differ about the exact character of the recurring cycles. So Pythagoras and his followers insisted upon the repetition of every detail; and Aristotle's pupil Endemus said of them: "If we are to believe them, you will sit in front of me again and I shall be talking to you and holding this stick, just as we are now." (An idea of exact recurrence that we shall meet again later.) Aristotle himself could declare that "the same opinions have arisen among men in cycles, not once, twice, not a few times, but infinitely often," though he had in mind only approximate repetitions, unlike Pythagoras. His actual analysis of Time, as we shall see, is acute; nevertheless, he does not deny, on the cosmic scale, its circular motion.

In that glorious morning of the ancient Greeks —and perhaps mankind has never had another to match it—this idea of recurrence probably gave them a confidence few men except saints or happy innocents have ever known since. All things from the greatest to the smallest, they must have felt, were in their places. Although in its movement Time showed only broken images of Eternity, one was not entirely separated from the other. Time had some touch of the eternal in it. One age might be worse than another, men knew sorrow and death, but Time could not whizz them into some future that might be unredeemable misery, the end of all good things. If that early Golden Age was best, then Time was already curving around to bring it back. If Xenophanes was right and men by searching found out better, then this improvement could never be lost, because again and again they would find out better.

Even if (as some Greeks believed) there was no gradual development, but cataclysms would occur, destroying all civilization so that men would have to begin all over again, they also believed that, as the wheel of Time turned, the lost civilization would rise once more. (Unlike us, in the mid-20th century, who first abolish the idea of Time returning, and then threaten ourselves with a man-made cataclysm that could abolish everything else.) And though only the gods were immortal and man must die, he and everybody and everything he loved had their own kind of immortality, just because Time instead of hurling them into oblivion for ever must curve around to bring them all back again. We shall see how, in the Roman twilight of the age, this idea of recurrence revealed another aspect of itself, but in this bright morning it gave men more confidence than we have ever known.

Though they might all have more or less the same cyclical cosmology as background to their thought, often accepting a complete cycle as the

Great Year, the Greeks were too intellectually curious to accept the notion of Time without examining it. As early as the sixth century B.C., Heraclitus, who left us those fragments about our never being able to go into the same river, must have pondered the problem of the persistence of identity despite changes in Time. (But we go into the same river not because the water is the same but because it flows between the same banks. These of course could be changed by time, but then a point could be passed after which we could not call it the same river.)

A little later, Parmenides, a notable influence, took up a position in which he was joined afterward by many metaphysicians. He denied the reality of sense-evidence, change, and Time. Like the Indians he declared that the world as revealed to the senses, the world in which Change and Time appear to exist, is an illusion. Behind this illusion, in the realm of reality, there is no "becoming" but only single and indivisible "being," changeless, timeless. Past and future are part of the illusion; what really exists is a perpetual present. Later philosophers, unless they rejected Parmenides, had no excuse to be idle, faced as they were with the task of reconciling his bold conclusions with the world around them, in which Change and Time certainly appeared to exist.

Among these later philosophers, and one who thought his master Parmenides unanswerable, was the ingenious and paradoxical Zeno, who must be given a high place among our world's intellectual entertainers. Believing that every metaphysical position except that of Parmenides must result in logical absurdities, he it was who gave us the runner who could never reach the other side of the stadium, the flying arrow that appeared at every given moment to be at rest, Achilles who could never overtake the tortoise.

The ruins of the great palace of Knossos—center of the Minoan civilization that flourished about 4000 years ago. From Crete to Rome, most of the ancient classical world conceived of Time's movement as cyclical—a cycle that would eventually bring the return of a golden age of civilization.

Zeno's paradoxes have often been argued and refuted, always by men better equipped than I am for the task; but if we keep to our subject Time and slyly introduce infinite divisibility, we could add another paradox and prove that a clock, having struck 12, could never arrive at five minutes past 12. It reaches, we will say, one minute past 12, then 61 seconds past, then 61 and a 10th of a second past, then 61 and a 10th and a 100th of a second, then 61 and a 10th and a 100th and a 1000th and so on. The clock is now, so to speak, moving inward rather than onward, just as Achilles cannot overtake the tortoise because they are no longer racing toward the goal but toward infinity. Or so it seems to me, but I am not addressing myself to mathematicians.

Plato refused to follow Parmenides, did not dismiss Time as an illusion, but accepted it as "the moving image of eternity." He acknow-ledged two forms of existence: Being, belonging to eternity, and Becoming, a characteristic of the natural world. What is revealed to our senses is an imperfect changing representation of an un-changing eternal model. It is the ordered regularity of Time that makes it possible for us to accept it as an image of Eternity.

Time, coming into existence with the universe, has reduced—or is reducing—chaos to order, making the motion of the universe harmonious and intelligible, bringing Becoming nearer to pure Being. The temporal world is a kind of compromise between pure Being and a meaning-less multiple Becoming. Time can be identified with the periodic movements of sun, moon, and planets, created "to distinguish and guard the numbers of time." He appears to have regarded Time and Space as being quite distinct from each other. While Time and the universe arrived

together, Space existed before they did. (But outside Time, what is "before"?) He would seem to have been less curious about Time and more anxious to create a perfect cosmology than some philosophers before and after him.

Aristotle, in his own cautious and common-sensical fashion, comes closer to the problem. He asks if Time exists, and, if it does, then what is its nature. One must suspect, he says, that either it does not exist at all or only barely and in an obscure way. For, taking a closer look, he points out that it is composed of the past, which has been and now is not, and the future, which is going to be but is not yet. What is made up of things that do not exist could only have a dubious share of reality.

What we know we have of Time is *Now*. Yet Time cannot be held to be made of *Nows*. A *Now* is not a part of Time, for a part is the measure of the whole. As the end of the past and the beginning of the future, *Now* is a kind of link. Time and motion, he concludes, are interrelated. Things not affected by the passage of Time must be outside it. As Space exists only in so far as there are bodies that occupy a certain place, so Time exists only in so far as there are bodies that at different *Nows* are in different places or states. Time is the number of motion according to "before" and "after," number being that which can be counted.

The recognition of Time involves a perception of *before* and *after* in motion, and a numbering process based on this *before* and *after*. Without some mind or soul to number it, there can be no Time. As the movement of the heavenly bodies provides the numbers of Time, Aristotle concludes that "If there were more heavens than one, the movement of any of them equally would be time, so that there would be many times at the same time." And for a moment we feel as if Einstein may be just around the corner.

Later, the Epicureans and Stoics, almost dividing philosophy between them, held opposing views of most things but shared some common ground in their ideas of Time. It was still moving around a vast circle: So the Stoic Apollodorus of Seleuca could declare that Time is the interval of movement of the cosmos, and

that whole time is passing just as we say that the year passes, on a larger circuit. But the opposing school, being atomists, would not accept the divisibility of Time; they believed in the existence of a present moment, for nothing could be infinitely divisible because all division must arrive ultimately at the indivisible atom.

To the Stoics, influenced by their concept of the continuum, what was thought of as the present was in fact partly future and partly past. One of their leaders, Chrysippus, said "No time is entirely present." And there gradually emerged the idea of Time, which we know only too well, as absolute, a universal inner flow unrelated to anything external. Even in the third century B.C. Strato could say "Day and night, a month and a year are not time nor parts of time, but they are light and darkness and the revolutions of the moon and the sun. Time, however, is a quantity in which all these are contained." In other words, the concept has now been abstracted from experience.

The idea may seem to have had a setback, for two centuries later we find Lucretius, the great Roman poet and a thoroughgoing atomist, declaring that Time "exists not by itself, but simply from the things which happen. . . . We must admit that no one feels time by itself abstracted from the motion and calm rest of things." But he must have been challenging a widely accepted opinion, which had followed Strato and many another in accepting Time as a universal container, existing whether we acknowledge it or not.

Ideas of the larger sort, keeping hold of men's minds for many centuries, have their dawns, high noons, sunsets, and twilights. In the Greco-Roman and then the Imperial Roman world, the idea of cyclical Time and the recurrence of all things arrived at its dusk. Not that it was no longer believed, but it was accepted in a different spirit. That eternal center of the gods, around which the ages revolved, lost its brightness and beauty. Where there had been immortal beings, who could be worshiped with joy, there were now iron Fate or capricious Fortune or the pitiless stars that "know in the midst of our laughter how that laughter will end." So everywhere in the huge Roman Empire, including the highest places from which it had to be governed, the astrologers multiplied, and far-reaching decisions depended on their reading of the stars. In fashionable Rome strange gods and goddesses appeared, to be tried for a season, like new styles of hairdressing or exotic dishes.

And these people were still not in History and rectilinear time, as we have long felt ourselves to be. Time for them still went endlessly round and round, but now like some blood-stained old wagon rumbling into the dark. The profound corruption of slavery; the organised sadism of the games and circuses; the sensuality that soon reached satiety; the murderous intrigues that passed for politics; the cynical cosmopolitanism of the huge city; the worship that had lost both heart and head and was dwindling into superstition: this was the Rome that was to turn at last, in hope or despair, to Christianity.

Here, in place of that confidence and inspiration felt long ago in the clear bright morning of Greece, we find—in Professor Puech's words—"a certain melancholy and weariness, a more or less intense feeling of anguish and servitude. This inflexible order, this time which repeats itself periodically without beginning or end or goal, now appears monotonous or crushing. Things are forever the same; history revolves round itself; our life is not unique; we have already come to life many times and may return again, endlessly in the course of perpetual cycles of reincarnations. . . ."

And what does Pater's Marius the Epicurean hear when he listens to the emperor Marcus Aurelius's discourse to the Senate? "I find that all things are now as they were in the days of our buried ancestors—all things sordid in their elements, trite by long usage, and yet ephemeral. How ridiculous, then, how like a countryman in town, is he, who wonders at aught. Doth the sameness, the repetition of the public shows, weary thee? Even do doth that likeness of events in the spectacle of the world. And so must it be with thee to the end. For the wheel of the world hath ever the same motion, upward and downward, from generation to generation. When, when, shall time give place to eternity?"

Re-reading it after many years, I have not rediscovered in *Marius the Epicurean* the book I admired in my youth. But as one phrase after another catches the eye—"a weight upon the spirits"; "the incurable insipidity even of what was most exquisite in the higher Roman life"; "an experience which came amid a deep sense of vacuity in life"—Pater does suggest an atmosphere of staleness, weariness, melancholy. Now it will be remembered that he takes Marius nearer and nearer to Christianity; the people who watch over Marius in his last illness are Christians; and it is their belief and their conduct that bring him to this conclusion: "There had been a permanent protest established in the world, a plea, a perpetual afterthought, which humanity henceforth would ever possess in reserve, against any wholly mechanical and disheartening theory of itself and its conditions."

Nevertheless, we have no feeling here of a gushing new spring of life, hope rising like a fountain. Somehow the staleness, weariness, melancholy, still remain: We are in the twilight of an age. And I cannot help feeling that Pater succeeds here, and fails to suggest the bright hope and wonder of early Christianity, just because he felt obscurely that he too, like those noble but weary Romans, was now living in a not dissimilar twilight, toward the end of another age.

Finding himself responsible for an Empire in danger of disintegration, and desperately in need of some binding force, Constantine finally brought Church and State together. By doing this we might say he put Rome and the Empire into History, the history that the rise of Christianity had created. The cycles of Time, now old and creaking, were stopped, then soon forgotten. Henceforward, Time moved along one straight line, taking the whole Western world toward some hidden and secret destination.

And though even 80 years ago Pater may have had some doubts, never spoken perhaps and only unconsciously revealed in his account of another age that was ending, officially we are still being hurried along that one straight road, belonging to History and to Time, wondering harder than Constantine and his Bishops, who set us all in motion, where we are going.

6   Time, History, and Eternity

A late 15th-century portrait of
St. Jerome (*c.* 340–420), one of
the original "Doctors of the
Church," as a medieval scholar.
During the early development of
Christianity, the idea of an eternal
Paradise seemed an escape from
earthly life and ordinary Time;
later, Eternity was conceived as an
endless continuation of passing
time—a concept that was to gain
prominence in medieval thought.

6 Time, History, and Eternity

1. Early Christian

In his magnificent large-scale study *The Uses of the Past*, Herbert Muller begins his section on the Mission of St. Paul as follows:

Among the teachers of the ancient world was one who taught that behind all the gods was a supreme deity, and that the highest goods were love and selflessness. His piety was such that he was reputed to be the son of a god, though he himself made no such claim; he performed miracles, casting out demons and raising a girl from the dead; and when he died, his followers maintained that he appeared to them afterwards, and then went bodily to heaven. The teacher I am referring to was not Jesus, however. He was Apollonius of Tyana, as described by Philostratus. He illustrates the lofty religious ideals against which the followers of Jesus often had to contend. He forces the question of why Christianity triumphed over its rival mystery religions, which are now discredited because they failed, but which also offered high hopes to man and at their best made high demands.

And in a footnote he points out the likeness in so many particulars—even in the rites of baptism and communion, the celebration of the sabbath on Sunday, an annual festival on December 25—between Mithraism and early Christianity.

One good reason why the latter was triumphant, Muller goes on to say, was that it found in Saul of Tarsus, later St. Paul, a missionary of genius. What he preached had little to do with the historic Jesus, a preacher of a different kind. His concern, so fierce and deep that he gave himself no rest, was with the resurrected Christ, the Savior-God, the Redeemer, who was revealed in that vision on the road to Damascus. A guilty man crying to other guilty men, in a world doomed to be stricken with sin and death, he preached salvation through Christ. He must have blazed through that pagan twilight like a rocket.

Though himself a Jew, Paul took this new and startling religion, as far removed from perfunctory offerings to Fortune as a cry of "Fire!" is from a bid at bridge, out of Judaism into the wide world of the Gentiles. It was for all men—women too, at a pinch, if they did not make too many female demands—because all men shared the original sin of our species and, however unworthy of God's grace, could seek and find salvation through Christ. He struck terror into bored pagans and then gave them hope. This new and intensely dramatic doctrine could neither be shrugged away by the philosophers nor stamped out by the police. "We preach Christ Crucified," the cry rang; and those who believed were willing, often eager, to be taken out of Time, whether this meant a few moments of agony in the glare of the circus or a long lingering death, and transported miraculously into the Eternity of Paradise. They had no doubt now that while they were in the body, before the soul was set free, they were firmly fixed in History and Time. But they also believed that History had only a brief course to run. And Paul himself had declared, with a fierce emphasis we can easily imagine: "The time is short."

Two ancient haunting ideas, which we have already considered, vanished with the triumph of Christianity. The first, itself the earlier, was the idea that profane passing time was nothing, impossible as it was for men to receive any sustenance of courage and strength from it, when compared with the sacred ever-enduring Great Time of myth and of gods and heroes. Jesus Christ did not belong to myth, and was not simply another of these gods or heroes, now mere creatures of fable. He was in one sense an historical personage, born at a certain place at a certain time (though the Church declared this date to be December 25 chiefly because this had long been accepted as a festival), and dying to this life at a certain place at a certain time. His earthly life was in History. Everybody and everything else, therefore, the whole life of man, the world itself until the day of its destruction, were there too. Passing time could not be negligible because this was the time *with which* (not *in which*) God had created the world. Time was switched on at the creation, to be switched off sooner or later, and sooner rather than later.

It is true that in the liturgies, which in one form or another went back to very early Christianity, communicants might know some eternal living present in which the supreme drama of death and resurrection was still being re-enacted; but there was no sense of a constant escape from meaningless passing time into the Great Time, the eternal dream time. What had happened historically now gave meaning to men's lives. The realm of myth, shining somewhere but supported by pillars that went deep into the dark of men's minds, was out of bounds.

The Church's acceptance of and dependence upon History and its rejection of myth have been severely criticized in our own age. Christianity built its foundation on the wrong level. It began to entangle itself in a world of stubborn facts. Even its miracles and marvels, acceptable and often delightful on the level of myth, had to be historical, to take place when and where they could never have taken place. (Some of Gibbon's sliest ironies are on this subject.) Sooner or later it would have to fight a rearguard action where it should never have been fighting at all. History, bristling with more and more facts, turned traitor to it. The very truths to which it could genuinely lay a claim were endangered, with some loss to the human spirit, because it denounced knowledge it could not afford to understand, and threatened with hell-fire good men with whom it compelled itself to be at cross purposes.

Centuries of bitter conflict, leaving wounds not healed yet, came out of this rejection of myth. Moreover, as some depth psychologists, notably Jung, have told us, a culture first shaped and colored by Christianity has reached us perilously lacking in that fructifying mythical element which thousands of generations of men found necessary for their imaginative life. This may mean that we are living without myths, empty in our depths, or that, as our ancestors' children, we are still myth-making but in a negative inferior way, gilding and lighting up any rubbish.

The second of the two ideas that vanished with the final triumph of Christianity was the idea of cyclical Time and recurrence. As we have already seen, though it still took its place among philosophies, it was accepted without joy or hope in the Roman world, where it seemed to create an atmosphere of weariness. Augustine singled it out for sharp condemnation. It was another attempt of the infidel, he announced, to undermine "our simple faith." Christians were not to be dragged from the straight road and compelled

A sixth-century mosaic of St. Paul whose vigorous evangelism spread Christianity all over the Mediterranean world. Christianity, uncompromisingly based on historical facts, uprooted two former ideas about Time: the concept of a Great Time distinct from passing time; and the classical view of Time as a recurring cycle.

to walk on the wheel. It was against the true faith to believe that the same revolutions of times and of temporal things were repeated, that when Plato sat in the Academy in Athens, teaching his pupils, he was doing something he had done over and over again through countless ages of the past and would repeat through countless ages of the future. God forbid we should swallow such nonsense! *Christ died, once and for all, for our sins.*

This was a unique event. It had happened once, here on this earth, in a certain place at a certain time, and any suggestion that it could happen over and over again was a whisper from the devil. Man does not come into existence through some constant shifting of atoms or some pattern-making by natural forces. He was created, an embodied soul living within an order both organic and spiritual, to avoid sin, to seek happiness, to glorify the Creator. The greatest sin (here Augustine agreed with other early theologians) was that pride, prompted by the devil, which encouraged the mind "to worship either itself or the body or its own vain imaginings."

Man was now in History, moving toward a future but not one that could be determined by science or metaphysical speculations; it was sacred, fulfilling prophecies, so that sooner or later it would involve the conversion of the Jews, the reign of Antichrist, the second coming of Christ, the last judgment, the conflagration and renewal of the world.

All of these, Augustine declared, we are bound to believe will certainly come to pass, but how and in what order is a matter of experience at the time rather than for the mind of man to apprehend fully at the present. And here we have a good example, on the part of an extraordinarily acute thinker, of that forced confusion of two quite different worlds—one belonging to myth and symbolism and the other belonging to historical fact—that at a later time would weaken the whole authority of the Church.

When he comes to consider Time itself, Augustine asked a question and followed it by an answer that is famous: "What then is time? If no one asks me, I know; if I want to explain it to a questioner, I do not know." But he knew enough to steer a difficult course between the meta-

physical rocks and shoals of his age. To some extent his conclusions are the result of a series of brilliant compromises. He rejected Aristotle's bracketing of Time and motion, and avoided any confusion of temporal with spatial concepts. Time was to be found in the soul or mind. There —"when we measure silences and say that a given period of silence has lasted as long as a given period of sound, we measure the sound mentally, as though we could actually hear it, and this enables us to estimate the duration of the periods of silence. . . . If a man wishes to utter a prolonged sound and decides beforehand how long he wants it to be, he allows this space of time to elapse in silence, commits it to memory, and then begins to utter the sound."

But this, he points out, involves a constant passage from the future to the past. "All the while, man's attentive mind, which is present, is relegating the future to the past. The past increases in proportion as the future diminishes, until the future is entirely absorbed and the whole becomes past." But of course the future does not yet exist and the past no longer exists. The mind, however, has three functions; "expectation—for the future, attention—for the present, memory—for the past."

This might be said to be the first appearance of the "common-sense" view of Time now generally held. But quite apart from the whole complicated theological structure that Augustine erected, with which we have nothing to do here, there is one important difference between his view and that of the first six persons we might stop in the street. He went inward—a shrewd decision, this—to discover Time, but he also believed that further inward still, where movement has ceased and all is quiet, is the true Eternity: not the roundabout of the cyclic cosmos but the timeless realm of God.

The pagan philosophers, he declares, "strive for the savor of Eternity, but their mind is still tossing about in the past and future movements of things, and is still vain. Who shall lay hold upon their mind and hold it still, that it may stand a little while glimpse the splendor of eternity which stands for ever: and compare it with time whose moments never stand, and see that it is not comparable." We are then, as embodied souls, within History and Time; but as souls, even while still in the body, we may in moments of contemplation or intense feeling (as he described in his account of the last hours he spent with his mother before her death) catch a glimpse "of that eternal Light," touch "that eternal Wisdom which abides over all."

Here he seems to return to that Jesus of Nazareth, now almost lost in the vast theological structure based on Christ and the *Logos*, who said "The kingdom of heaven is within." And a host of Christian mystics, in century after century, can be said to have followed him. But it could be argued (see, for example, Alan Watts's *Myth and Ritual in Christianity*) that the mischief had been done. (I am now taking, of course, a long-term view.) By exchanging myth and symbol for history and fact, by making Time real in terms of expectation, attention, and memory, Christianity condemned itself to become an *onward* rather than an *inward* religion. It had to work on the wrong level. History began to contain it. Time changed it. Facing hostile criticism, it clutched at a rationality that did not belong to it; and by rejecting myth and symbol, it sealed off springs of awe and wonder and joy on its journey across the desert of history.

This was long after Augustine had died and the Roman Empire had fallen into ruin. He and the other mighty bishops and Fathers of the Chruch might claim 1000 years for the domination of their ideas, which contrived the most flexible and subtle—and perhaps the most difficult to interpret in terms of a common way of life—of all the great religions. In our age, we are told, Christianity has not failed; it has simply not been tried. But it must be held partly, if not chiefly, responsible for a climate in which it cannot be tried. And long ago, I repeat, the mischief was done.

2. The Maya

We will now escape from our own Dark Ages to the other side of the world, where the Maya were flourishing. They deserve a place here because they were infatuated with Time. They were obsessed by it as no other people have been. But first a few facts about them. What is called the Classic period of their civilization lasted, roughly, from A.D. 300 to 900. It covered an area that now includes part of Southern Mexico, Yucatan, Honduras, and Guatemala. Its staple diet was based on maize, easily and abundantly cultivated at that time, and this meant that for some centuries it was not difficult to find labor for immense public works, roads that often served no practical purposes, and huge complicated structures composed of pyramids, temples, terraces, and courts. (In their ball-courts the Maya used balls of solid rubber, 1000 years before we thought of such things.) Their civilization was not destroyed by conquest; possibly the soil began to show signs of exhaustion; and it is thought that the peasants, having to work harder and harder to support a formidable ecclesiastical hierarchy, finally revolted. Perhaps they felt that that obsession with Time, up there in the temples, had gone far enough.

The Maya seem to have been the mildest of the Middle American folk, very different from the Aztecs, whose final orgies of human sacrifice suggest they were more than half out of their minds. Any blood demanded by the Mayan gods appears to have been freely offered by devotees, human sacrifice being comparatively rare. Their sculpture, murals, pottery, jade ornaments, are ample proof of their sensitiveness and skill in the visual arts. (Though I will confess here that I can never escape the feeling that their style of drawing, like that of other Middle American peoples, has something crazy about it, reminding me too often of drawings done by our own schizophrenics.)

But their whole civilization was curiously out of balance. They made astonishing advances in some directions and never moved an inch in others. They could make roads that were engineering feats, and could build on a most impressive scale. (The so-called "Acropolis" at Copan was erected on a platform covering 12 acres.) And, as we shall see, their calculations, done without mathematical aids, and their astronomical observations, without the benefit of instruments, were astonishing. Yet they had no domesticated animals, no wheeled vehicles, and only worked in metal at all toward the end of their civilization. They were a very odd people, but if seemingly a little mad, not unpleasantly so. Fortunately, unlike other American cultures, the Maya developed hieroglyphic writing, mostly surviving on *stelae* (limestone shafts), and though not all the texts have been completely deciphered yet, we know a good deal about these people, especially in terms of their fantastic obsession with Time.

Although mention is made of "cycles" and "eternity" by historians of the Maya, it seems to me that their idea of Time was strictly rectilinear and that "eternity" here simply means unending passing time. The Maya regarded all divisions of time as burdens carried by relays of divine carriers. Some of these were benevolent, others malevolent. It was very important, there-

fore, to determine who would be "carrying the day." There was repetition, it is true, because one set of Time-carriers would succeed another in this curious cosmic relay race, and in this sense the future might be like the past. But this seems to me something quite different from the Greek idea of cyclic Time and eternal recurrence.

On the contrary, the Mayan idea is not far removed from our own concept of Time—a one-dimensional track going straight from the past into the future. And if we feel like dismissing with contempt their belief in complicated alternations of "good times" and "bad times," born of the feeling that they were at the mercy of benevolent or malevolent Time-carriers, we might remember that modern man finds himself equally at the mercy of strange economics, bringing booms and "recessions," about which he feels he knows even less than the Maya did about their Time-carrying gods. For it was the task of the Maya priesthood to calculate, by methods that became increasingly complicated, how this relay race of good and evil was being run.

Left, hieroglyphs for the nine known Mayan units of time. The Maya evolved a very complicated calendar, involving both a secular and a ceremonial year. They were also remarkable artists, though much of their art—for example, the incense burner (c. 1400) below—suggests (as does their obsession with vast computations of time) a slightly unbalanced quality in their civilization.

All the intellectual life of the Maya went into these calculations, so elaborate that I refuse to be drawn into them. This paragraph from Eric S. Thompson's *Rise and Fall of Maya Civilization* illustrates how complicated they came to be:

The general problem that faced the Maya might be expressed in terms of our modern civilization in this way: In the speedometer of your motor car you have a mileage gauge, which also records tenths of a mile. Suppose you have installed also gauges which record the distance you drive in furlongs and chains, in kilometres, in Roman stadia, in Russian versts, in Spanish leagues, and other measures of distance. Now suppose only the mileage gauge is illuminated at night, and, furthermore, that you, a very superstitious driver, fear that you will be in danger if several of the dials show your unlucky number, five, at the same time. On the other hand, when the gauge registers a group of sevens, your luck is in. You must calculate when the fives will appear on the gauges during your night journey so that you can drive with great caution; and also discover when the sevens will be in position, so that you can speed to make up time lost during the dangerous period.

He goes on to point out that the problems the Maya priests had to solve were in fact considerably more difficult than this, based as they were on solar, lunar, and planetary observations and calculations. Moreover, various other periods of time, sacred and secular, had to be taken into account. And they had no instruments for their astronomical observations, no mathematical aids for their calculations, no short cuts, not even any way of handling fractions or a decimal system.

Within its limitations this Maya priesthood (so many arithmetical astronomers and calendar makers and astrologers working in close co-operation for centuries) must be credited with some astonishing achievements. They brought to their calendars an accuracy and precision not known in Europe until long after their time. They worked on a staggering scale; one of their computations goes back 400,000,000 years. In most matters they appear to have been an exceptionally mild, calm, orderly people; they were only immoderate about Time. No doubt the priests who spent their lives making these calculations believed they were doing an important public service—the more that was known of the relay runners of Time and all their combinations, the easier it would be to foretell the future—but I suspect that these Maya intellectuals, forming a separate and highly integrated community, soon became hypnotically fascinated by their calculations down long vistas of Time.

It is just possible that some future age, discovering our passion for science and technology, our willingness to follow wherever they might lead, might see us in the same light as we see this Maya priesthood—earnest, devoted, clever people but perhaps not quite right in the head. It is also possible—for intellectuals sometimes take wrong turnings—that, in their passion for surveying this one-way track going further and further into the past and the future, the priests might have come to have a less rewarding idea of Time than the peasants whom they instructed.

I put forward this possibility, rather tentatively, because of a passage in Helen Augur's *Zapotec*, a sympathetic study of the Indians who live to this day not far from where the Maya once raised their pyramids and temples. She points out that "we are all vividly conscious of time in the sense of having so much of it to spend before the next appointment and before the final appointment with death," and that "to us time is horizontal and in movement," "a one-way current which inevitably carries us toward death," and how "this conception is intolerable; we all secretly hate and fear time." All this is in sharp contrast, she declares, to the Zapotec's idea of Time:

To the Indians, time is vertical, and it does not move "somewhere." Time cannot move because it is also space—an idea very difficult for us to grasp. It gives the Indian a relation to the three tenses that is different from ours. We live in a very narrow present. . . . The Indian lives in a present which contains all the time there is. To the Zapotec the past is not something behind him, but around him. . . . The modern Zapotec has held on to the general attitude which was the heart of the classic religion. Instead of fearing time he worships it as a divine gift. Because he is mortal there is fear in his attitude towards the future, but this is fear in the religious sense. The Old Zapotecs believed that time shared human mortality and that only the kindness of the gods could keep it going. The Middle American clock stopped dead at the end of the fifty-two year calendar round.

Here I must interpolate that the Maya combined a 260-day religious year with the 365-day civil year to give a period of 18,980 days—or 52 years. Helen Augur continues:

Every half-century the Middle Americans faced the possibility of universal death. . . . They believed that by prayer and sacrifice they must keep their gods kindly disposed; and at the crucial moment of decision all mankind must pray together for its clock to start ticking again. The choice was total extinction or total worship. . . . Each individual was impelled to live aright in the sight of the gods, for the life of all humanity depended on him. It seems a big burden for one soul to carry, but the core of Indian thought is that the individual is the race and vice versa, and thus all burdens and blessings are shared.

Now some of this probably came down from the Maya, including of course the choice of a 52-year period. But it must have survived on the level of unlettered folk belief. These Indians had not been deciphering the hieroglyphics of the Maya sacerdotal astronomer-astrologers, many of them half-buried for centuries in the jungle. What had long outlived Maya civilization were certain popular beliefs about Time as described by Helen Augur. But these seem to me very different from the idea of Time on which the vast Maya researches were based. Behind the centuries of observation and tireless calculation was the idea of Time proceeding in a straight line for millions and millions of years. The divine bearers might carry it in all manner of different relays, bringing men good, ill, or indifferent fortune, but they were all hurrying one way.

It seems to me, therefore, that to these many generations of Maya intellectuals—and this is what the priests were—Time was also "horizontal and in movement." The idea of a 52-year bottle of Time, so to speak, giving the Zapotec a relation to the three tenses different from ours, releasing him from our "very narrow present," may have filtered down from the Maya, but by way of the popular imagination and not, I suspect, as a legacy from its priesthood, that early example of conscientious intellectuals engaged in teamwork. This Zapotec attitude to Time, as Helen Augur describes it, seems to me much closer to that of primitive man.

I may be all wrong about this—leaving myself open to be corrected by any authority on the Maya—but the point I am trying to make, no matter how wrong or right I may be about the Maya, is that the intellect, used analytically, sooner or later reveals an idea of Time very different from that of men whose minds remain on a "mythopeic," imaginative, intuitive level. I am suggesting that in this matter the ignorant and unlettered may in their way come closer to the truth than learned and brilliant calculators.

Sharp analysis and precision may trap us in a dead end. People who keep their minds open, never refusing what intuition and imagination may bring, are frequently dismissed as being "woolly-minded." There is in my opinion much to be said in favor of being woolly-minded. No doubt all kinds of nonsense may find their way in and be hospitably entertained, but so might a profound truth or two. Such a truth may easily be considered irrational, illogical, paradoxical, not worth a serious examination, by the opposite of the "woolly," the sharp and keenly analytical minds, determined to cram the universe into the limited human intellect and equally determined to ignore whatever has been left outside.

So it is, I feel, with this subject of Time. As soon as it is examined and analyzed by the intellect, without benefit of intuition and imagination, it shrinks to a concept that in the long run, as we shall see, is not even intellectually satisfying, and that leaves our whole mind, if we still disregard the promptings of intuition and imagination, both bewildered and appalled. I am not one of those who deny any thought of essential progress; we move on and up and in spite of many setbacks, perhaps more than most world historians will admit, we make some solid gains; but on a wide and generally accepted level (to which the very newest theories and researches have no access) I cannot see we have made any progress in our ideas of Time. There may have been more rewarding beliefs in the cave and the jungle than there are now in the office and the street. But to this I shall return later.

A Mayan "almanac" stela erected in A.D. 497. The elaborately carved figure represents a sky god, who presides over the five-year period of time inaugurated by the stela; other carvings record the dates and predictions of important events.

3. Medieval

After Augustine and the crumbling away of the vast Empire came the Dark Ages, and out of the Dark Ages came the Middle Ages. But it seems to me that when we are considering ideas of Time, and hoping to discover a single line of development from the rise of Christianity to our own day, we can easily go wrong about the Middle Ages. I cannot escape the feeling that they cannot be seen as a bridge between Augustine's world and that, say, of Newton. (I mention him here because he is important in the next section of this chapter.) There is a sense in which Augustine and Newton were closer than either of them would have been to a typical medieval character.

If we see these ideas of Time moving along a highway, then I believe we have to turn down a side road to reach the medieval world: Somewhere around a corner we shall find the castle, the knights and men in gay livery, the walls and smoke of some town above which rises an astonishing cathedral. In spite of its universities and disputating schoolmen, its wandering scholars and monks meditating in their cells, in spite of the centuries that separate it from St. Augustine and other saintly theologians, it has not gone deeper into Time and History. It might almost be said to have come out of them. If that is too strong, let us say that the Middle Ages, which were not really in the middle of anything, had a peculiar character all their own, developed well away from the highway of Time and History. And my guess is that today we are more like Roman citizens, pagan or Christian, than we are like our medieval ancestors.

In the Dark Ages many peoples and cultures, Latin and Germanic, pagan and Christian, barbarous and civilized, vigorous and effete, were thoroughly mixed, and out of the new mixture, coming to the boil, came the medieval world. It was in many ways a narrow world, and probably for that reason all the more intensely felt, but contained within itself the most violent contradictions. The play of the opposites generated tremendous energy. Medieval man found himself involved in a huge drama of good and evil. Angels and demons battled for his soul. Above the small sunlit stage was the blue of Heaven, below it the fires of Hell. He swung dizzily between the grossest sensuality and the pure flame of the spirit. His leaders might order a massacre on Tuesday, feast all Wednesday, and go barefoot in penance on Thursday. He could live in a hovel and help to build cathedrals that still take our breath away. He might see on one hand all that war, famine, and pestilence could do to men, and, on the other, pageantry more sumptuous than anything we have ever seen. His was an elaborately ordered society, in which every man had his place, but everything in it rushed to extremes. It was a world with a heated imagination. So it was enormously exciting to itself. The grey boredom in which millions of us exist today was regarded then as a special sin. For all our technological marvels, we would probably seem machine-like and dim, half like robots, half like ghosts, compared with our medieval ancestors.

If that is thought to be far-fetched, here is a test of its truth. Read again Chaucer's account of his pilgrims, each so sharply differentiated, so distinctly and solidly himself or herself; and then take a good look, either in real life or in fiction, at any group of contemporary travelers, waiting for some agency's chartered airplane or motorcoach. After that we can ask ourselves what it was that brought Chaucer's pilgrims, representing so many levels of medieval society, into one traveling company. Religious belief plus custom? Perhaps. But bearing in mind our subject, the ideas of Time, I suggest a different answer.

What brought them together and then enabled them to ignore status and social backgrounds was certainly not any feeling of being in Time together. We are far more conscious of ourselves as the victims of Time than they were, and this if anything encourages us to cling to and assert our social status. No, I feel that what these medieval characters had in common was concerned not with Time but with Eternity. It seemed, we might say, *very close to them.* This sense of its closeness—the human scene being more than half illuminated by its white radiances or its hellish flames—appears to be profoundly characteristic of the Middle Ages. It belongs to that age and to no other, before or since.

Both the human scene and the cosmos, it seems to me, closed in during the Dark Ages and then expanded again at the Renaissance. It was one thing to live in the Roman Empire, even when it was about to disintegrate, and quite another thing to live in and with the Holy Roman Empire, a kind of wild fiction, being, as Voltaire said afterward, not holy, nor Roman, nor an empire. I feel that History and Time on the one hand and Eternity on the other were on a different and altogether broader and larger scale to, say, Augustine and his contemporaries. The medieval world was narrower, more dramatic, more intensely felt. (In spite of the fact that it contained more bold free thought than it is generally assumed to have had.)

It is true that the end of the world and the Day of Judgment, which earlier Christians felt could not long be delayed, receded as the centuries passed; but medieval historical ideas were as vague and semi-magical as its notions of geography, in which every blank space was filled with fantastic monsters. Though every university might be noisy with dialectical argument, some of it extremely subtle, the air trembled with magic, entrancing or sinister. Medieval man gave his imagination free play.

There is an idea to which romantic literature has often paid a wistful tribute, and it has come to us chiefly from the Middle Ages, which were fascinated by it. This is the idea of the magic place out of Time, really a kind of outpost of Eternity in this world. It may be deep inside some forest or at the end of some forgotten winding road. It may be behind that door in the wall which we see only at certain moments. In *The Romance of the Rose* it is a walled park; and now I quote C. S. Lewis's account of it from his masterpiece, *The Allegory of Love*:

... If we could pass the wall we should see the white flocks, who are not shorn or slaughtered any more, following their Shepherd through the good pastures. The turf there is covered with little flowers which the sheep browse on, but whose number never decreases and whose freshness never withers. For all this region is out of the reach of time. Never does Jean de Meun show himself a truer poet than when he touches on the idea of eternity—of all metaphysical ideas the

richest in poetic suggestion. If the pleasure I take in these lines is "philosophical," then so is my pleasure in *Kilmeny* and in folk tales and morning dreams innumerable.

For alwet stant in oo moment
Hir dai that nevere night hath shent,
Ne hit conquered nevere a del;
For no mesure temporel
Hit nath, but alwey as at prime
That brighte dai, withouten time
Of temps futur ne preterit,
Also hit laugheth, evere abit.

And it is this walled garden, apparently, which we were really in search of all the time. . . .

One reason why so many of us, at one period of our life or another, have fallen under the spell of the Arthurian legends is that they seem to wander in and out of Time. The magic place, some Avalon where there is never rain nor snow, some castle of good or evil enchantment, may be around any corner. It is this atmosphere of wonder rather than any obvious picturesqueness that drew the romantics to the Middle Ages.

It is not that these people were not aware of Time, which often made a dramatic appearance: "Shall not I, Time, dystroye bothe se and lande/ The sonne and mone and the sterres alle?" But I cannot help feeling that this Time—not strictly medieval, for this author wrote just after the 15th-century breakdown of the Middle Ages—does not really reach the sun and moon and all the stars: He is almost a local rather than a cosmological figure, a sinister old acquaintance of the family. I cannot believe that medieval man ever felt trapped in some vast machinery of Time. Time was at work but on a small and homely scale, not flowing ceaselessly through unimaginable space. History, in which imagination filled any gaps, was almost on a similar scale, and hardly worth bothering about. As Coulton tells us, the Second Coming

... never seemed far beyond the immediate horizon. Treatises on Antichrist abound. Even Roger Bacon, writing in 1271, speaks of the common belief among "wise men" that this last stage of the world is imminent. Dante leaves very few seats to be filled in his Paradise; and, if he could be recalled to earth, he would probably be less surprised at any modern invention, than that the world had lasted 600 years after his death. The same belief meets us again in almost every generation; Sir Thomas More himself was inclined to it. Therefore, while the best medieval thought was deeply serious, and penetrated with the sense of personal responsibility, yet it sometimes suffered from an impatience which was the defect of these qualities; the world-fabric might crash at any moment; what was the use of painfully beginning a long and continuous chain of facts and inferences which involved the labour of whole generations or centuries, when a few years or even weeks might bring the consummation of all things? . . .

So the future was dubious. As for the past, whereas Augustine and his fellow bishops looked back to the Empire and the classical tradition, the medieval scholars looked back to the Dark Ages.

The figure of Time, in the allegory of Stephen Hawes quoted in the last paragraph, arrives only to bring in the more triumphant figure of Eternity. The idea of Eternity no longer suggests something remote from this world. Like the magic timeless place, it is almost around the corner. Everything, so to speak, has now closed in. The classical age that had gone and the modern age that was to come took and still take what might be described as an immensely broad horizontal view of human life, in which the individual is easily lost sight of. But the medieval view is narrow and vertical, like that of a stage brightly illuminated on several levels, a stage for the drama of the individual soul. (A drama too in which elements of the grossest farce and the blackest tragedy were bewilderingly mixed.)

It is as easy to despise or condemn the Middle Ages as it is to offer them too much sentimental admiration. But within limitations impossible for us to accept, they achieved something that somehow or other, on our much broader base, we shall have to match in our own way if we wish to escape radioactive ruin or a world-wide ant-hill—and that is an ordered society, sustained by a common and not ignoble belief, in which the individual was not lost but discovered himself, not suppressing but releasing joyous energies, reaching to the sky in the shape of cathedral towers, for the glory of God, not in towers of office blocks for the worship of Mammon. I declared earlier that the single track of Time, the highway

straight to our age, could be said to by-pass the medieval people or make a curve around them. We should begin to look for a curve or two now.

4. Enlightenment
This is the best place in which to make a point of some importance, one that seems obvious as soon as it is made. It is this: In any society at any given time, people exist consciously (their unconscious is a different matter) on various mental levels; they are never all believing or thinking the same thing. Even in our modern society, with its universal elementary education, its newspapers, radio, and television, there are still people who feel and behave almost as primitive man did, others who are more or less medieval, others again who seem to be still in the 18th century.

We tend quite rightly to associate an age with its newest and most original ideas, and there is no harm in this as long as we remember that only a few men, at that time, may have actually held those ideas, and that many decades, often amounting to centuries, may pass before those ideas have seeped down to wider and commoner levels of belief, thought, and feeling. So many "hard-headed and realistic" men of today repeat what scientists were saying nearly 100 years ago, and may know nothing about the outlook and prevailing moods of scientists today. And men in the street now assert beliefs originally found among the intellectuals of the 18th century. We must expect time lags of various lengths.

It was Whitehead who pointed out that we are still living upon the accumulated capital of ideas provided "by the genius of the seventeenth century." It was then that Newton wrote that "Absolute, true and mathematical time, of itself, and from its own nature, flows equably without relation to anything external." Now this concept was challenged and denied over and over again, notably by Leibniz, who argued that Time was not absolute but relative, that "instants apart from things are nothing." Leibniz was followed by other metaphysicians, mostly allowing no reality to Time, right down to our own age. But no matter how high and wide their fame (that of Kant, for example) these metaphysicians never broke through into the broad vague realm of popular belief. They did not succeed in creating an acceptable world picture. They left the general climate of opinion unchanged.

It was Newton—though he may never have been studied or even properly understood—who was believed. His largest bequest to the public mind was not his general theory of the physical universe, no matter how important that was to succeeding generations of physicists and astronomers, but his concept of Time.

(There is a story of Newton worth repeating. He had been badgered by an old woman, who believed him to be an astrologer and a magician and so begged him to tell her where she could find a ring she had lost. Finally, weary of being pestered, he told her the first thing that came into his head, simply to get rid of her. She could find the ring, he said, if she went so many paces down a certain street. And that, we are told, is exactly where she *did* find the ring. If, for the sake of argument, we accept this story as truth, then we have here two quite different worlds, Newton's and the old woman's; and the old woman's, magic and all, seems nearer the real world, in which, however, otherwise the ring would never have been found, Newton was living too. Most scientists would have us believe that they, and we, exist only in the world in front of their eyes, where things may be examined and tested, and not at all in the world behind their eyes, in which so much cannot be examined and tested. And they do not help us to find any rings we have lost.)

No 18th-century English poet was quoted more often than Pope, and it was he who wrote:

Nature and Nature's laws lay hid in night:
God said "Let Newton be!" and all was light.

And as I have said elsewhere, in *Literature and Western Man*, it would be difficult to find two other lines of verse that tell us more about the 18th century than these do. Notice their air of confidence and cheerful pride; their suggestion that at last all is known; their hint that God can now retire, Nature and Nature's laws and Newton taking over; their final emphasis upon *light*. Darkness, with all that was unknown, mysterious, superhuman, magical, fateful, has vanished. Reason, the inquiring and experimenting mind, brings everything into the light. So very soon all the enlightened, if they allowed God still to exist, regarded Him as a remote First Cause, who had put together and then set in motion the vast but smoothly running well-oiled machine of the universe. God did not intervene in human affairs; worship and prayer were wasted efforts; men should study Nature and use their Reason.

And in this comfortable Deism, like an armchair, the 18th-century intellectuals took their ease. They began to take an interest in the past; the bookshops were filled with new historical works, including the masterpiece by Gibbon; and most of these historians saw the past as a tangle of overgrown paths arriving at last in the broad avenue of the present age. Time, as Newton had said, might flow "equably without relation to anything external" but it had brought powdered wigs, satin waistcoats, and snuff-boxes, civilization and the triumph of Reason. A gentleman with a taste for philosophy might have read Leibniz on Time, Berkeley on our dependence on sensations and our ignorance of matter, Hume's skepticism about mind and causality, and yet feel that Newton's Time was carrying him along in sensible and friendly fashion.

What I want to suggest now is that while everything else changed, this idea of Time remained. The scientists' Nature began to show disturbing traces of femininity; she refused to show the same face and figure to them; she revealed evidence that demanded new theories to explain it; she would not settle down with Reason and the inquiring mind. Complacency was shattered by sharp reactions and challenges in religion, social theory, politics, literature: evangelical movements, mystery-mongering circles, the French Revolution, the Reign of Terror, Napoleon, the Romantic Revival. But whatever else went, the Newtonian idea of Time survived.

It was there to greet the 19th century as it had greeted the 18th. So far as there was any change in it at all, it might be said to have been strengthened and increasingly sharpened. Time's flow had something more urgent about it, as if it were starting to run downhill. Men began to feel it was carrying them along now in no sensible and friendly fashion. The repeated shocks of the Revolutionary and Napoleonic wars encouraged this feeling. But there was another and far more important reason why men began to have a different feeling about Time.

This can be found not in the French Revolution but in the Industrial Revolution. The new inventions and disciplines took man deeper into Time, made him far more aware of its passing than he had ever been before. The old easy natural rhythms of work disappeared. The steam engine was outside the varying seasons, did not recognize any difference between night and day. It imposed its own rhythm on the men who worked for it. Any dreamy relationship with it had to be severely discouraged. The little towns that were built around enormous mills—and many of them exist today, almost unchanged—were inhabited by people among whom only the very youngest children, mere babies, were unaware of the relentless times of the machines. Everybody else was at their mercy.

And though the humblest workers, including for many generations even young children, may have felt that an idea of Time was being imposed upon them, there soon came into existence almost a new race of men, content to work not *for* but *with* the machines. These inventors, engineers, mechanics, could not avoid, even if they had wished to do so, a new relationship with and consciousness of passing time. And soon they brought millions of other men—and their more reluctant women, for there is something in Woman that hated it, and still does—closer and closer to this consciousness, this relationship.

An obvious example of this change is not hard to find. The Industrial Revolution had been at work for at least half a century when railway trains began to take the place of stage coaches. Nevertheless, the coaches for the most part still existed outside the new atmosphere of sharpened times. They kept easy hours, leaving in the early morning, about noon, or toward sunset. Minutes did not exist for them. And this was not good enough for the railway trains. They left at 8.35 or 10.27 precisely, soon making their passengers —to say nothing of their drivers, stokers, guards, signalmen, and porters—realize that hours can be divided into minutes, and that each minute may be important.

And ever since then, with increasingly accurate time instruments at our command or commanding us, we have been compelled to become more and more aware of smaller and smaller divisions of time. (How often have I heard, when recording for radio, somebody say "We'll begin—in ten seconds from *now*." And there are still hundreds of millions of people in the world who would not be able to understand that sharpened *Now*.) In the advanced science and technology of our own age, even a second can be regarded as a great clumsy piece of time, so that it is divided by a thousand—into *milliseconds*. Now it could be argued that when a *millisecond* means something, when we are aware of smaller and smaller time divisions, we ought to be expanding and enriching our experience; and, for all I know, this may be true for the eager experimenters who play with *milliseconds*. It is not true yet for all the people outside labs.

In the 19th century the rapid growth of industry broke up ancient patterns of living, put an end to tradition, packed its workers into dense urban areas, and created at least an image of a grey faceless horde—"the masses." This movement from the country, with its old seasonal rhythms, to the industrial towns, with their dark, narrow streets and factories like fortresses, their engines and machines, their hooters and whistles and clocks, made men increasingly and disturbingly aware of Time. And after two or three generations, as religious belief decayed, the idea of Eternity lost all meaning for them.

In this respect they were far worse off than the primitive men to whom missionaries were being sent. For them there was no Great Time always there, waiting for them. They had no rites and ceremonies that might admit them into the eternal dream time. There was no *other kind of time* into which they might escape. Everything that could ever happen would have to happen along that single track. The kingdom of heaven could not be within—*and now*. It had to be in the future. And it is from industrialized Western Europe in the 19th century that our social and political-economic utopias have come. These must not be confused with ideas for reform; they have a different feeling and tone; they could be described as an attempt to restore the eternal dream time, but not to its own dimension. It has to go into ours, further down the line, into the future.

The most successful of these attempts, as we know now, was that of Karl Marx. What brought him success in the end, long after his death, was that unconsciously he endowed his communism with elements of myth, essential to the eternal dream time. The drama behind his figures and dialectical arguments is an ancient one of gods and heroes. Professor Eliade has pointed this out, declaring that Marx followed one of the great myths of the Middle Eastern and Mediterranean world—the redemptive part to be played by the Just or the Elect (now the proletariat), whose sufferings in the end will save the world. His classless society, after the triumph of the proletariat, belongs to the myth of the Golden Age which, many traditions asserted, lies at the beginning and the end of History.

"Marx," says Professor Eliade, "has enriched this venerable myth with a truly messianic Judaeo-Christian ideology; on the one hand, by the prophetic and soteriological function he ascribes to the proletariat; and, on the other, by the final struggle between Good and Evil, which may well be compared with the apocalyptic conflict between Christ and Antichrist, ending in the decisive victory of the former. It is indeed significant that Marx turns to his own account the Judaeo-Christian eschatological hope of an *absolute goal of History. . . .*"

We can add that it is also significant that he has failed just where he thought he would first succeed, in highly industrialized Western Europe. He has probably won most converts through the unconscious, irrational, mythical elements in his grand design. They promised to take man out of History, and to restore—but now bringing it down to earth—the Great Time. It is the strength of communism that these mythical elements can capture the imagination, especially among people no longer held by their own traditional myths. It is a weakness of communism that once the imagination frees itself, once these mythical elements lose their magic, the atmosphere in which communism lives begins to seem boring, stifling, sterile.

The 19th century accepted the Newtonian Time of the 18th century, but I suggest—and I am not being merely fanciful—the universal flow was speeded up. Time, like man, was moving faster in the 19th century. The 18th-century intellectuals had seen mankind arriving at last at civilization, Reason, an understanding of Nature's Laws, an inevitable broadening freedom from superstition, bigotry, political and ecclesiastical tyranny; these were bringing man where he belonged. There was much tidying up to be done, of course, but intellectually and emotionally all was settled; there was no hurry; Time could move easily. The difference between this attitude of mind and that of the 19th century is the difference between the sound of a trotting horse and the roar of a train. Everything was moving faster, including the sense of passing time.

This helped to produce the idea of constant progress, a world getting better and better year after year, as most politicians, scientists, historians, newspaper editors, were ready to testify; and the opposing idea, held by some philosophers, social theorists, and critics, and several of the age's most important novelists, that everything was getting worse and worse and that the world was running into appalling danger. Even the optimists, ready to believe that progress was now automatic, felt they were living in and with a time that moved faster than it had seemed to do to 18th-century thinkers. As for the pessimists, the deniers of progress, from Carlyle to Dostoevski, they often make us feel that Time was roaring in their ears as it raced toward catastrophe.

Extremes of optimism and pessimism, progress and regress; the claims of novelty and tradition, what was too new and what was too old; science challenging religion, Socialism and Communism denouncing Capitalism, realism opposing romanticism: Out of these opposites and the battles between them came that extraordinary energy of the 19th century and all those great seminal ideas that we in our century inherited and, instead of driving them out with ideas of our own, proceeded to put into practice on a large and often catastrophic scale.

And in this huge but creative turbulence, the idea of Time, linear, inescapable, queerly ac-

celerated (just as Evolution had been accelerated), played a considerable part. The clock
struck oftener and ticked louder.

There were many attempts, on different levels
of intelligence and feeling, to escape. Schopen-
hauer almost returned to the vast life-denying
outlook of ancient India; Nietzsche offered a
new version of the old Greek idea of Eternal Recurrence; Bergson, reversing the usual process,
dived into Time to discover a semi-magical
duration. The Transcendentalists could hardly
write five lines without mentioning Eternity.
With the continuing help of St. Augustine and
St. Thomas Aquinas, Rome embraced new
converts, from Oxford and many another place,
and as a bonus produced Papal Infallibility.

Romanticism, strong in the first half of the
century but weakening later, was among other
things an attempt to escape this idea of Time, to
return to those magical places of the Middle
Ages. Spiritualism, beginning as a fashionable
fad about the middle of the century, soon included among its devotees several distinguished
scientists. True, it seemed to accept without
question a linear one-dimensional Time, common to those on this side or the other side of the
grave; but it was one form of protest against Time
as a conveyor belt carrying us to extinction.
Then Madame Blavatsky and Annie Besant gave
the later 19th century the Theosophical movement, which introduced into innumerable tea-
shops and parlors Reincarnation, Karma, Astral
Planes, and various lost continents.

These were the years when Myers and Gurney
were the pioneers of psychical research and published the first and perhaps the best book of its
kind, *Phantasms of the Living*. Then in the later
1880s and early 1890s, Hinton and several other
speculative mathematicians were responsible for
a curious little flurry of books about the fourth
dimension. Hinton was a genuine original who
has not had the attention he deserves; in one of
his awkward attempts at storytelling, *An Un-
finished Communication*, he surprisingly anticipates
Ouspensky's theory of recurrence. This is surprising because Hinton and his fellow fourth-
dimension theorists had not associated this unknown dimension with Time. This was left to a
young man of genius, Herbert George Wells, a
former science instructor still in his twenties, who
in 1895 published *The Time Machine*.

It will at least serve as a bridge between the last
century and this one if I add a note here about
H. G. Wells. I was fairly well acquainted with
him during the last 20 years of his life, and my
admiration and affection for him are on record
in more than one place. Thinking about him
now, with the ideas of Time still in mind, I find
myself regretting deeply, for his sake as well as
ours, his later attitude toward the Time problem.

He must have deliberately turned away from
it, no doubt believing that other things were more
important, but I always felt that this decision had
left him feeling uneasy. He was irritable when
the subject came up. He pooh-poohed Dunne,
whom he had known long before *An Experiment
With Time*; he dismissed with a snort Ouspensky
and his followers, without knowing much about
them; and although he was never rude about it,
he deplored the way in which I was bothering
my head about Time in the '30s. He was like a
man who, having wrongly given up playing an
instrument for which he had a flair, then refused
to listen to anybody else playing it.

And I feel strongly that if Wells, after writing
his *Outline of History*, had dedicated himself to
solving the riddle of Time, he would have saved
himself much disappointment later, he would
have rediscovered his astonishing gift of intuition,
he would have left us, at this date, far deeper in
his debt. His last thin cry of despair, *Mind at the
End of its Tether*, must be taken for what it is, the
work of a dying man, unconsciously projecting
his own dissolution upon the universe. But I
think he might never have written it if, instead of
ceaselessly pleading for "open conspiracies" and
scientific world governments, he had brought all
his great gifts to the Sphinx of Time.

After all, it was a certain idea of Time, which
he might have challenged and then destroyed,
that encouraged the age to reject his open conspiracies and plans for a sensible world government. One reason why important people would
not listen to him was that they felt they were
being hurried along to their graves and wanted
to grab as much as they could on the way there.

5. India and the East

This section on Eastern ideas of Time, coming as it does just before my account of This Age, may seem to be in the wrong place. After all, these Eastern ideas are anything but new; they go back nearly 3000 years. But while they are old, they cannot be relegated to a distant past; they are still widely accepted in the East, and indeed have not been without influence on certain kinds of Western thought. Therefore it seemed sensible to delay my account of them until I had reached this point, at which I can bring to them the emphasis they fully deserve.

Although it was over 50 years ago, I have not forgotten the moment when, after exploring the maze of Indian metaphysics, I reached its central thought. I read that if we go deeper and deeper into the self we can arrive at last at the recognition of Atman, the essential self; and that if we go deeper and deeper into the not-self, the world that seems so solid and real, pulling aside veil after veil of illusion, we shall find Brahman, the ultimate reality; and that Atman and Brahman are identical. I can remember the exultation I felt, being then in my late teens, when if we have any sympathy at all with speculative thought we long to encounter bold and gigantic metaphysical conclusions.

Perhaps I was offered then one of those "favorable moments"—the discovery of the door in the wall where none can be seen—by means of which, if the opportunity is seized, the Indians believe we can begin to escape from the bondage of Time. I did not seize the opportunity, was bound again to the wheel, and perhaps this is one reason why I am writing this book. Time, to which I returned myself, began to haunt me, as an idea, just because I might once have had the chance of escaping from it.

We seem, however, to be guided almost from the first by certain natural sympathies. So almost everything in traditional China, its philosophies and arts and style of life, its essential earthiness and sly humor, appeals to and at once gains my sympathy; while almost everything in traditional India, its huge fine-spun metaphysical webs, the vacuum in much of its spirituality, its gods and goddesses with too much of everything so that they look more like freaks than divinities, leaves me feeling indifferent or repelled.

I am neither defending this attitude of mind nor apologizing for it, but simply stating a fact. Traditional Indian life and thought are rejected by a certain skeptical, coarse-minded, determinedly occidental element in me. I cannot help entertaining some doubts about all this enormous spirituality, telling myself coarsely how pleasant and welcome it would be, there in a hot country, with too many people, too much noise and dust, too much life bursting out all over the place, to retire to a green shade, close tired eyes, and conclude that this world is an illusion. But having admitted so much, slapping my cards on the table, I must also admit that in any account of man's ideas of Time, India must be given a prominent place. Its speculative thought has been Time-haunted. Time is the villain in its huge cosmological drama.

That there is something fascinating about the part it plays, I do not deny, and indeed could not deny after quoting earlier with such approval that story of Narada and Vishnu. But then this idea of the relativity of Time, implying different degrees of illusion and reality, and the notion that what may seem half a lifetime to one mind may appear to another as a minute or two, may have a special appeal to me just because I am a storyteller and a dramatist. On the other hand, whatever makes such a special appeal to a storyteller or a dramatist may, without his being aware of it consciously, be a myth that symbolizes a profound truth, to which he may be unconsciously drawn. Then he may struggle against recognizing this truth, and his very reluctance to do so may create conscious prejudices against a way of life and thought in which this truth is embodied. I am, I hope it will be agreed, trying hard to be fair all round.

Believing that the mind or soul of man should be weaned from any belief in Time, Indian thought has adopted drastic methods. It first sickens and then terrifies us by saying in effect: "If it's Time you want, then you shall now have more than a bellyful." The idea of cyclic cosmological Time had not been unknown to other and earlier peoples, but the Indians methodized and elaborated it on a terrifying scale. So, briefly, to avoid wandering in a wilderness of strange names and astronomical figures: One day for Brahma, the period of existence of a universe, and one night, during which this universe arrives at dissolution, are each equal to 4,320,000,000 human years; and these daily creations and dissolutions of universes will continue for "a hundred years of Brahma," a mind-reeling total of human years that the reader must work out for himself.

And even after these thousands of billions of our years Time will not come to a stop, for when Brahma has completed his task, other gods will rise to follow, and more and more universes, with their unnumbered worlds, will come into existence and vanish in their turn. This is cosmology on a scale that matches the discoveries of astronomy, offering us whole galaxies by the million.

Tales are better than figures. There is a myth (which I am borrowing from Professor Mircea Eliade's rewarding essay on *Indian Symbolisms of Time and Eternity*) that dramatically illustrates the stupendous scale of this Indian cosmology. Indra, the victorious King of the Gods, was becoming too self-important and needed correction. Disguised as a ragged boy, the supreme deity, Vishnu, visited Indra, and began to tell him about the innumerable Indras who had already existed in the innumerable universes: "Like delicate boats they float on the fathomless, pure waters that form the body of Vishnu. Out of every hair-pore of that body a universe bubbles and breaks. Will you presume to count them? Will you number the gods in all those worlds—the worlds present and the worlds past?"

At this moment a procession of ants, four yards wide, begins to cross the floor of Indra's palace. The boy-Vishnu sees them, and breaks off his talk, first in astonishment and then to burst into laughter. Indra asks him what he is laughing at. The boy-Vishnu replies: "I saw the ants, O Indra, filing in long parade. Each was once an Indra. Like you, each by virtue of pious deeds once ascended to the rank of a king of the gods. But now, through many rebirths, each has become again an ant. This army is an army of former Indras. . . ." And the vanity, pride, ambition of Indra withered away.

There is here of course a double-edged attack upon our belief in ourselves in Time, upon our confidence that we can find our reward in Time. First there is this factor of sheer scale. Against these unimaginably vast circlings of Time, our portion of it, together with our importance, dwindles alarmingly, shrinks to a speck.

You have been elected, let us say, chairman of the company, which is what you always wanted to be. The company is important, now you are important, and you can hardly resist swelling and strutting. But just before you are about to take the chair for the first time in the board room, a proud event you have long wished for and may have secretly rehearsed, an old emaciated Indian magically appears in your office. He begins by pointing out that a single daytime of Brahma lasts 4,320,000,000 of our years. Queerly unable to order him out of your office, you reply that you are not interested in this improbable figure of years. Ignoring this, the Indian begins explaining the time cycles within the great cycle, and how the *mahayuga* consists of four ages each of them longer than your historical records, and that 1000 *mahayugas* constitute a *kalpa*, and that 14 *kalpas* make up one *manvantara*; and as he drones on, you cannot help having a distressing vision of everything you know and understand shrinking and shrinking—your success, the company, the country you try to serve, the whole civilization itself, the very world, the life of man, all dwindle and empty themselves and become tiny bubbles that flash, wink, and are gone. As your visitor vanishes, you are left bewildered, wondering, humble. (And quite possibly might make a better chairman.) And this, we may say, is the Indian Time Trick.

A note here. I have been looking again at Radhakrishnan's *The Philosophy of the Upaniṣads*, for many years now a standard work, and I feel that I must warn the reader that much of what I say here, not always with the gravity the subject demands, would be sternly rejected by this distinguished Indian philosopher. Bringing ancient Indian thought almost into line with Western idealist philosophies—almost as if he had given the authors of the Upaniṣads a few terms at Oxford, round about 1900—he denies the common view that we have here a belief that what we accept as reality is an illusion, merely the conjuring tricks of maya, and that Time as we know it is a cheat, and that we must see through it to save ourselves. Thus:

Brahman is infinite not in a sense that it excludes, but in the sense that it is the ground of all finites. It is the eternal not in the sense that it is something back beyond all time, as though there were two states temporal and eternal, one of which superseded the other, but that it is the timeless reality of all things in time. The absolute is neither the infinite nor the finite, the self or its realisation, but is the real including and transcending the self and its realisation, life and its expression.

And again:

There has been much criticism of the theory of the Upanisads under the false impression that it supports the illusory nature of the world. It is contended that progress is unreal because progress is change, and change is unreal since time in which change occurs is unreal. But the whole charge is due to a misconception. It is true that the absolute is not in time, while time is in the absolute. Within the absolute we have real growth, creative evolution. The temporal process is an actual process, for reality manifests itself in and through and by means of the temporal changes.

This shows us a refinement of traditional Indian thought, not without some traces of Western influence. But I feel it still leaves me free —working here as I must in a rough-and-ready fashion—to take from this vast and complicated tradition a line of thought that not unfairly represents Indian ideas of Time. In short—and, as politicians and senior civil servants like to say, with all due respect—I must bypass Dr. Radhakrishnan, saluting him as I turn aside.

Now we reduced that proud new chairman of the company to a proper humility chiefly by contrasting his miserable portion of time with gigantic eons in which the total existence of his company, or indeed almost anything with which he could identify himself, is too small to be seen. But this idea of scale is only one edge, and the less important edge, of the double-edged attack I mentioned. And already the other edge, cutting deeper, is at work. This is the idea that in time

there can be no sense of completeness, no enduring satisfaction. Even the gigantic eons themselves are doomed to vanish. Everything in time, no matter what its scale, is here today and gone tomorrow. We can cry "So what?" to the eons and the universe-bubbles just as we can—and often do—to our own pitiful helping of years. As Yeats wrote in his magnificent old age:

All his happier dreams came true—
A small old house, wife, daughter, son,
Grounds where plum and cabbage grew,
Poets and Wits about him drew;
"What then?" sang Plato's ghost. "What then?"

"The work is done," grown old he thought,
According to my boyish plan;
"Let the fools rage, I swerved in naught,
Something to perfection brought";
But louder sang that ghost, "What then?"

Because we feel ourselves to be completely contained within a one-way temporal process, looking straight ahead like the driver of a car, we tell ourselves we must work for our children and children's children, we must serve the future. But where along this conveyor belt is supreme value to be found? What can we reply to that ghost? Why should our children's children be able to find the right answer? Where in the future will it be found?

We are busy, as they say, "passing the buck." Pinning our faith to evolution, more than content with some Great Chain of Being, we may see ourselves as so many cells in what will eventually be a glorious Tree of humanity. But this could only grow and bloom outside of Time, which could only condemn it to wither and rot and finally disappear and be forgotten. If we discover wonder-working drugs to increase our span of years, we only give ourselves more time in which to brood over the final victory of Time. If we take ourselves to other planets, we take our insoluble problem with us; under two suns and five moons we are still doomed to a final darkness.

This second edge cuts deeper because it is concerned not merely with quantity but with quality. No matter how the empirical self adapts itself to the concept of passing time, a one-way horizontal track, the essential self (which expects something different and better) tries to escape from the contradictions, the ruthless opposites, and knows nothing but a sense of frustration, a profound dissatisfaction. What are our philosophies, when they are something more than logic-chopping, and our arts, when they are not revelations of inner chaos, but attempts of this essential self to escape, to cry a halt to passing time? And is it not when that self begins to feel itself being finally defeated that philosophy dwindles to logic-chopping and the arts offer us revelations of inner chaos, denying us hope when we most need it? As for religions, all religions that are more than systems of formal morality and empty ceremonies, have they not rejected the empirical self and taken their stand on the essential self, while rejecting its frustration, its dissatisfaction, taking it out of Time?

Let us take the first crude example that offers itself: that in passing time we always seem to be various and succeeding sections of ourselves, never our whole selves. This at least is no freak of self-tormenting Indian thought, anxious to wither the world away. It is an idea that runs its melancholy course through the poetry and popular sayings of the West. It can be found in our verse from Shakespeare to Ella Wheeler Wilcox. (There is too much to quote from him, but we can give Ella Wheeler her chance: *If I were a man and a young man*/*And knew what I know today*.) It haunts all those wry conclusions—*Si jeunesse savoit, si vieillesse pouvoit* and the like—that belong both to literature and to folk wisdom. We may be "wanting better bread than can be made of wheat," may be like children crying for the moon (but not the one we may shortly be offered), we may be refusing to come to terms with reality, but it cannot be denied that this longing to realize, in our relationships and actions, our whole selves is deeply felt, and can be discovered in some of our noblest minds.

We cannot help but feel that Time as we generally understand it cheats us, that we have a self that cannot take part in its charades. We seem compelled to wear a series of disguises. When we are young, we are young—and something else. When we are old, we are old—and something else. And the something else, we feel,

is more important than youth or age. There is an essential John Smith who is neither young Jack Smith nor old Johnny Smith. These are, in a sense, a pair of impostors. They and others like them occupy the stage in one scene after another, while the true whole John Smith waits vainly in the wings until the curtain comes down.

It is not a mere egoism in us that makes us feel we are being cheated. We may believe (as many intelligent and sensitive people do nowadays) that, so far as we have a main task here, it has nothing to do with the "saving" of individual souls, and that we exist to add experience, knowledge, a broadening and deepening consciousness, to some kind of *anima mundi*. All this, and more that encloses us within a great tree of mankind, may be believed, yet the frustration, the dissatisfaction, could still be deeply felt.

And not irrationally, for we could argue that these successive sections of ourselves, never adequately representing the whole self, fail to gather experience and knowledge of the highest value, and that whatever it is that needs them cannot accept as a substitute our feelings of dissatisfaction and frustration. Therefore, though we may shrug away any idea of saving our souls, time-bound we cannot begin to fulfill this apparently more impersonal task. And indeed, time-bound we shall find it hard to conceive of any kind of *anima mundi* to serve, for it cannot be seen as a remote terminus on the one-way track of passing time, cannot be confined to the fourth dimension, must have existence outside our time.

It has been argued that all this is nothing more than a childish refusal to accept reality. We have been conditioned by thousands of years of irrational religious beliefs, myths, fairy tales, and now, in an age dominated by science and technology, we must wake up and face the truth. Passing time is all we have, and we must make the best of it. All this longing to escape from its limitations and contradictions is no better than wanting to have one's cake and yet eat it. It is desiring the impossible, and so should be rejected firmly by any fully mature intelligence. The popular mind should now be thoroughly reconditioned to accept reality, to live contained within the concept of irreversible and inescapable entropy time. And indeed something like this is happening, especially among our new urban industrial people, brought closer to the second law of thermodynamics than their country cousins.

And with what result? Upon many of them there has descended a curious apathy, a new boredom, a lack of zest, a flavorless sense of living, from all of which, we must remember, the scientist or technologist may be protected by his curiosity, his sense of power, his vanity. Others, especially among the young, react differently, though with no more gain to the community, because they are at the mercy of mysterious feelings of frustration and anger, turning to violence out of some irrational conviction that they are being cheated. So the new children of Time tend to sink into dejection or scream with rage. But we must not give up hoping. The science that has closed one door may soon find itself opening another. Time—at least our prevailing idea of it—has not said the last word.

Though this is generally believed in the West, it is not true that traditional Indian thought demands an utter rejection of the world and Time. Even among the various forms of Yoga—and one of their intense self-disciplines, we are told, can bring about a changed body-mind relation with Time—there is one form, Karma-Yoga, that does not take a man away from action in the world and any familiar style of life. But he who practices it must not identify himself with the fruits of his action, must never lose himself in the world and its time, must be constantly aware of another and enduring reality. (*All that is seen is temporary*, says the Tamil proverb.) He may not have to believe that everything around him is part of an illusion, a web of deceit for which there seems little justification (God as an illusionist or designer of obstacle-race courses is not an imposing figure); but he must utterly reject the idea that this alone is reality, all there is, the sealed container of our species. His obligations may include much else; but this is enough.

Not being familiar with the East, I have not to my knowledge ever met anybody who was consciously practicing Karma-Yoga; but even with such prejudices as I have already admitted,

I feel that here are an outlook and attitude of mind that could be as valuable in Baltimore or Birmingham as they could in Bombay. It is the kind of detachment we often find in men who have accomplished great things while refusing to be devoured by events and Time. What these men do proceeds from a self that does not waste away; it appears to have some secret source of nourishment. They behave as if they believed, paradoxically, that in this world everything is important and nothing is important. Perhaps they know they are on their way out of Time.

One final idea is worth noticing, if only because it appears again in some bold speculations of our own age. In traditional and semi-mythical Buddhist thought, there is the idea of a perfected self, the Great Soul, the Buddha whose existence is outside Time. (This idea can be found elsewhere, in various esoteric traditions haunted by the notion of a self-created immortality, sometimes involving a body of finer and indestructible matter surviving death.) Now while this infinitely superior being can exist outside Time, in some blissful eternal Present, inspired by his compassion for ignorance and suffering he can also enter Time at will. He has been awarded, we might say, the freedom of the fourth dimension. And for him Time is reversible. That he can go into the past involves the idea that the past is still existing in its own time, and that brings us to the further and more bewildering idea that if these superior beings have such powers, it might be possible for them to begin changing the past. But any consideration of these notions belongs to Part Three.

The final point I wish to make, after offering no more than a few glimpses of a huge body of thought, is that Indian ideas of Time, apparently far removed from our own, generally held to be utterly alien to Western thinking and feeling, can in fact be brought much closer to us. They cannot penetrate the ideas we wear like armor, but they can be brought closer to the way we live with ourselves, uncertain and uneasy, behind that armor. The Indian essential self outside Time is related to that self of ours which feels, as scores of familiar quotations can testify, that it can never be fully revealed in passing time.

7 This Age

A ruined alarm clock in a burned-out house: a suitable symbol for the narrow, meaningless equation of Time with clock time, passing time, that is so closely linked with modern man's loss of value and meaning in his life.

7   This Age

I was walking down Lexington Avenue, in New York City, and glanced at a florist's window. Then I stopped to take a closer look at its brilliant display of flowers. There was something about them that did not seem quite right. I saw then the small notice, which said that all the flowers in that window were made of plastic. This tiny incident made a deep impression upon me. In that window, I felt, the modern world revealed itself in all its strange sad confusion.

To begin with, an apparent rationality barely concealed a depth of unreason and absurdity. On the most superficial level there is something to be said in favor of these plastic imitations of flowers. Unless they are closely examined, smelled and touched, they seem like real flowers; but unlike real flowers they do not droop and wither and die after a few days; they will last for weeks, months, perhaps years. But then they are not flowers. Nor are they—and this would be excusable—decorative objects taking the place of flowers and perhaps carrying a suggestion of them, as a kind of tribute to a vanishing world. They are pieces of a lifeless synthetic substance pretending to be flowers. They are intended for people who want flowers so long as they are not really flowers. We are now groping in that depth of unreason and absurdity already mentioned, a place where, whatever else we may lack, we shall not lack company. It is where too many of our contemporaries try to live.

I suggest we think about real flowers for a moment or two. They are alive, as we are alive, but of course in their own fragile, brief, and exquisite fashion. Within a few days they may bloom, fade, and vanish. And how unbearably poignant this would be if it happened only once in our lifetimes, if the freesias now scenting the air in this study would soon be gone for ever! (I may be told here that those particular freesias *will* be gone for ever, to which I can only retort that I do not know them as individuals.) But as the seasons return, so do the flowers. It is

this, just as much as the beauty, fragility, and brevity of their appearances, that enables them to keep so strong a hold on our affection and imagination.

I am writing here just after the longest and hardest winter Britain has known for a century, yet already the aconites and snowdrops and the smallest yellow crocuses are blooming, having conquered the iron ground, and our hearts go out to them. It is as if the flowers go out of Time, renew some pattern and principle of growth in Eternity, then appear again in Time. And with this rhythm of life and death and eternal renewal, plastics shaped and colored like flowers have nothing to do. They are dead things, so to speak, abstracted out of life. They are also typical products of our day and age.

It is within the concept of Time, abstracted not only from our experience of succession but also from a hypothesis of cosmic mechanism, that most people now feel that they exist. I have pointed out already how ideas seep down to wider and commoner levels of intelligence and feeling until at last they are believed to be solid realities. It is precisely the "hard-headed and realistic" who all too often exist in cages made out of largely discredited hypotheses. They serve prison sentences behind walls and doors they only imagine are there. Our last idea of Time is now a way of life—to use a phrase that only came into common use after most people could neither find a way nor enjoy much life. This idea has in fact changed its character and quality. That is because on the broadest level of humanity, to which it found its way at last, no idea exists as such, to be accepted or rejected purely intellectually; it is transformed into an emotionally-charged belief, a world picture, a guide to living. The result may easily be discovered in the talk of people in general.

Time may not be mentioned. But they have only one life to live, they say, and everybody only has it once and can't start over again, and we had better make the best of it while we can, and if you've any sense you'll make sure you get your share, here and now, because you can't take it with you, can you? These are made as statements of fact, though often given a defiant or defensive tone, as if some invisible grandmother, who used to believe all kinds of nonsense, were being argued with.

The talk may go further afield than this, but if you listen to it carefully—as I have often done with this theme in mind—you will soon discover how much of it can be referred back to this idea of Time transformed into a belief. Though the Great Times, eternal-dream-times, recurrences, and eternities may still haunt the dark of the mind, this consciously held belief rejects any thought of them. We exist for so many years and then "when our number's up" we are dead and buried. We belong entirely to passing time. Any experience that suggests we can ever escape from it is just fanciful and, if repeated and insisted upon, calls for a doctor. Notions about some other kind of time are daft. What does Science say? And Science knows, doesn't it?

Out of this belief something strange has emerged. It is indeed one of the unique features of our own age. The very people who believe that

Time will inevitably kill them are also the very people who insist upon "killing time." They are slaves who murder their master every evening. To put it another way, what should be most precious to them, for they believe it is all they have, is shoveled away as if it were muck. And if it tends to seem like muck, that is because value has departed from it. Now I know that many factors are at work here, most of them outside our subject, and I do not want to suggest I am losing all sense of proportion, but I venture to risk this: that one reason why a sense of value has gone is that a concept, itself not concerned with value, has been transformed into an emotionally charged belief that should have value. And so indeed it has—but negative value. Life contained entirely by passing time, a meaningless portion of years, is not felt to be more precious than ever, it is not felt to be precious at all. Positive value has drained out of it.

A dozen proofs of this, so far as it affects people in the mass, could be found in the nearest newspaper. I prefer, however, to offer as an example a certain attitude of mind that puzzled me for several years. As an active protester against nuclear weapons, I could not understand why people in the mass, in all the countries concerned,

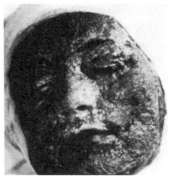

The obsessive quest for youth is a symptom of our feeling that we live entirely contained in passing time. Left, the interior of a beauty salon. Above, a more painful method of rejuvenation: Facial skin is burned off to produce a "new" face (right).

were so strangely apathetic. These people would not join us, to denounce the Bomb, yet they never made any effort to defend it. They were neither for it nor against it. They gave the subject a blank stare and then walked away.

Now I simply cannot imagine people in other centuries behaving in this fashion. Placed in such appalling jeopardy without being consulted, they would have hooted and stoned a parliamentary government out of office, dragged a king off his throne. Nowadays, it is true, we have mass media and expert propaganda to spread suspicion and fear. But the people I mean—and they form the great majority—are not suspicious and fearful, as many educated and more influential persons are. Propaganda has not made them accept the Bomb. We protesters, though we may have won over some of their sons and daughters, brothers and sisters, have not made them reject it. They remain profoundly, astonishingly, shockingly *indifferent*.

They are indifferent, I can only conclude, because in their heart of hearts they no longer care what happens. Consciously they may care about a lot of little things, especially in their chief capacity as consumers, but on the level immediately below consciousness and on the scale of the biggest things, they no longer care at all. If the other people, the supposed enemy, take over, well, that's that. If by some final piece of imbecility, the Bombs are dropped, so what? Let it all go, it was never really worth having. By using even catch phrases of this sort, I am probably falsifying the situation; it is not what comes into consciousness to be spoken but what is obscurely but profoundly felt below that explains this extraordinary indifference. And now it is worth repeating that life contained entirely by passing time, a meaningless portion of years, is not felt to be more precious than ever, it is not felt to be precious at all. Positive value has departed from it.

This inner nihilism, with its curious anesthesia, is responsible for one of the strongest features of our age. This is the tremendous and disproportionate emphasis, which modern advertising uses with terrific effect, first upon sex and secondly upon family life. The contemporary crowd does not consist of men more highly sexed than their great-grandfathers or capable of greater devotion to family life. Sex and the family are hammered away at, plugged, boosted, given the full possible treatment because they still represent some value where so much value and so much meaningfulness have vanished. If the bed and the home are all that is left, then it is the bed and the home that must be spotlighted, crooned over, praised by all those warm false voices coming out of microphones. Men and women have been making love and raising families for thousands and thousands of years; there is nothing new in this; what is new is the disturbingly empty space around these objects and activities of value.

And this empty space is really empty Time. It goes on and on, simply toward a future that these very people refuse to consider, but it cannot go in and in, toward the hidden springs and fountains of life. Unrefreshed by these springs, trapped in a barren concept of Time, these family men, for the most part, cannot even summon up enough energy and purpose to sweep aside the lunatics who threaten their children with Doomsday.

At first sight nobody would seem further removed from these inarticulate and unsophisticated crowds than the writers and the artists of the avant-garde, nothing if not highly articulate and sophisticated. But here, once again, the path curves round and extremes meet. The avant-garde literature and drama of the Absurd expresses to some extent the attitude of mind of the indifferent crowds. This is not the aim of avant-garde writers, who see themselves as an elite and make no conscious attempt to cater to mass audiences. And these audiences, when they are killing the time that is all they have, do not want their own inner despair or bitter resignation reflected back at them; they like their entertainment to be concerned, even if only lightly, with values and beliefs they themselves have lost; it makes, as they say, a nice change.

If you are living, deep within yourself, a "black comedy," you may insist upon relaxing with comedy that is as "shining white" as the clothes in detergent advertisements. The writers of "black comedy" and the Absurd are not altogether despairing or they would never trouble to write at all; something pleases them, if it is only a sense of their own cleverness. And much of this writing, though its authors may sharply disagree, is really a disguised protest: It expresses a belief in nothing but the Absurd but the writer has a hope, a kind of hidden half-belief, that this is an appearance behind which there is—or there ought to be—a reality that is quite different.

An intelligent and sensitive man who genuinely believed that life was entirely without meaning and purpose, an idiotic accident, would never take pains not only to write but also to negotiate and argue with agents, publishers, theatre managers, and the rest: He would commit suicide. The men in the mass do not commit suicide, though they probably do believe that life is entirely without meaning and purpose, an idiotic accident; but they are not as a rule intelligent and sensitive and, anyhow, generally do not exert themselves as much as the writers do or take the same risk of having their vanity wounded. And though they do not actually commit suicide, surely their strange apathy and their indifference to the threat of nuclear destruction have something suicidal about them? Men without this taint are active and determined protesters: They know that life itself demands that the Bomb must be rejected, and that to accept it, to live in its shadow, is to be indifferent to true value and to watch the flower of life being withered.

Erich Kahler, who was enthusiastically praised by Thomas Mann and Einstein, is not primarily concerned with Time in his wide-ranging study of modern man, *The Tower and the Abyss*. But he observes, in his notes on various great modern poets, "a profound feeling of contraction not only of space but of time—a gathering of all times and their contents, of our entire existence in one sublime moment, a concentration which is almost equivalent to an abolition of time." And in his later examination of the "stream of consciousness" techniques so characteristic of modern fiction, he turns to Time once more. These new techniques, he tells us, having broken through the bottom of our consciousness, have likewise "cracked the supposedly solid foundation of chronological time. A new time begins to germinate within time, the time of inner experience within the time of outer happenings. This new kind of time has no definite limits—the depths into which it expands are practically infinite. It cannot be measured by means of chronological time." Again, discussing Camus (whose huge reputation based on such a small body of work suggests that he has been accepted as a representative man of our age), Kahler cannot avoid a reference to Time: "Camus notes the sudden awareness of time that befalls us, the panicky feeling that we are tied to time. We constantly live toward the future which we should dread because it is death in disguise. . . ."

Kahler, I repeat, was not primarily concerned with Time in this study—its main subject is the effect upon human values of the powerful collectives, scientific, technological, commercial, of Western civilization—but the three quotations from him above seem to me important in our context. We might call them three separate bombardments, from different directions, of the great central Fortress.

This Fortress represents the greatest weight of authority known to us today. It is from here, not from Rome or any other religious center, that the iron dogmas and decrees come. It is the citadel of science, technology, positivism. If we feel we are tied to chronological time, that is where the tying was done. If we appeal against this servitude, that is where our appeal will be rejected. It is the home of science as a dogmatic system and a colossal vested interest. It is where most of the work, including the invention and development of nuclear weapons, is done. (How many maimed guinea pigs, cancerous rats, blind mice, and infected frogs exist within these walls, we cannot imagine.) Its platforms and turrets are manned and grimly guarded by a gigantic army of theorists, researchers, experimenters, professors, teachers, technologists, publicists, and journalists. Above the keep there flies the black-and-white banner of positivism.

But not all science and scientists are within these walls. The most advanced research, probably based originally on intuition and imagination, is not done here. And it is doubtful if there is one really great scientist in the place. It does not house any Institute of Advanced Studies. It makes no astonishing discoveries. This is where old hypotheses go when their best work is done, when they might die if left outside the ramparts. And not only do the dogmas and decrees come from here, but with them, perhaps far more dangerous, all the assumptions that can be found, far and wide, in schoolrooms, newspaper offices, debating societies, bars and cafés, and bedrooms in which husbands are telling wives not to be fanciful and silly.

It is from this Fortress that our climate of opinion has been controlled. Greater scientists, working in distant outposts of advanced research, are too busy, too deeply interested in what they are doing, and probably too open-minded, to work at this climate-rigging. They do little recruiting, too, except for their immediate purposes. But beneath the black-and-white banner, recruits are sworn in weekly by the thousand. Many of them come now not from the exact sciences but from anthropology, archaeology, psychology, sociology; they swarm into the Fortress, eager to dedicate themselves to measuring, weighing, graphs and statistics, determined never to make a statement not supported by the lab's figures; they insist upon being more scientific than their colleagues in the exact sciences. No more nonsense about humane studies, with a flavor of art about them; no more guesswork and intuiting; into the labs go one and all! Everything that can possibly be measured is being measured, and anything that cannot be measured deserves to be ignored.

This is not a mere gibe. Certain kinds of experience, for example, cannot be scientifically tested, if only because they may not survive the conditions of such testing. Thus we may believe in telepathy, but that does not mean we are ready to make an appointment at the Psychology Department's lab to prove it. And because we cannot be telepathic to order, that does not mean we have been bamboozled by a few coincidences. And it is no use the positivist telling us we are making meaningless statements, be-

cause what is meaningless to him may be deeply meaningful to us, and, after all, our real life is inside us, not inside him. The danger of this test-in-the-lab outlook is that it encourages the assumption that anything untestable by scientific method—falling in love, for instance—is unreal.

This danger (but in reverse, so to speak—that is, mistaking the unreal for the real) attends any over-dependence upon the statistical method. It can easily be forgotten that the average arrived at belongs to its own world of figures and not to reality. There is no actual citizen of Coketown who is 66.57 inches tall and weighs 149.6 pounds, is married to a wife 63.2 inches tall weighing 126.35 pounds, has 2.5 children, and earns £958 a year. I am not saying that such statistics are useless—though I may secretly believe that too much time, trouble, and money is now spent on them—but I do say they must not be mistaken for a closer and closer approach to real people in the real world. As Jung points out: "The real picture consists of nothing but exceptions to the rule" and so "Absolute reality has predominantly the character of *irregularity*."

The trouble is that the statistical method has behind it the immense prestige of the exact sciences. And now workers in other fields, belonging to what are at best inexact sciences, being anxious for research grants and the approval of distinguished elderly scientists, anxious too to be in the movement and not to have anybody in the senior common room sneering at

Left, the surrealist artist Salvador Dali (in a distorted photograph); above, a scene from Jean Genet's "theatre of the Absurd" play *The Blacks*; right, a young Japanese "beatnik" artist. Avant-garde art's eccentricity may be a reaction against the emptiness of modern life and a restricting idea of Time.

them, borrow as much as they can of this prestige by imitating the methods of the exact sciences. With the ironical result that very often now they are far narrower in their outlook, far less flexible and open-minded, than the younger and most advanced researchers in the exact sciences. With the further and still more ironical result that electrons and protons and neutrons seem to have more freedom than we have.

We may appear to have wandered too far from Time, but now we are back. Much of this new experimenting in psychology is concerned with one aspect or another of Time. Thus I have in front of me Gustav Jahoda's account, reprinted from the *Educational Review*, of the results of innumerable experiments to discover the development of the sense of Time and History in young children. The author of this admirable paper on "Children's Concepts," to whom I am grateful, declares in a letter that while not sharing my general views, he entirely agrees with me that Time "is a great deal more complex than most people imagine." But here, I think, he is referring to the difficulty young children have in fully understanding our concept of Time and realizing its various implications. What he does not mean is that this concept itself may be inadequate, and that the reality from which it has been abstracted may indeed be "a great deal more complex than most people imagine"—a very different matter. In other words, his stand is taken from well within the Fortress with the black-and-white banner.

There are many references to psychological and neurophysiological experiments and theories in the chapters on Individual Time in G. J. Whitrow's *Natural Philosophy Of Time*, a most comprehensive and excellent study within the limits it sets itself. It is the best book of its kind I know. It is a Fortress job, but as honest as it is thorough. The following paragraph, concluding the chapter on *Time and Physiology of Memory*, is worth reading, not only because it says something important but also because it is a good example of Fortress phraseology and manner:

In the present state of knowledge we are, therefore, obliged to conclude that, although both automatic and conditioned reflexes may be purely 'mechanistic' feed-back circuits, and although immediate memories may also be maintained by processes analogous to the dynamical circulation of memory in large computers, none of the many ingenious theories so far devised has succeeded in showing how our capacity for long-term memory can be explained in either mechanical or chemical terms.

There is indeed something very odd here. The brain registers and files away everything it has observed. Even the black-and-white-banner shock troops admit this, because experiment, either with hypnosis or stimulation through electrodes, has proved it to be true. Without warnings the files are opened, and there can be total recall, moment by moment.

I have known this happen without hypnotism or electrical stimulation of the brain, when alcohol apparently did the trick. I came back to an hotel late one night, and a man there, drunk but far from being speechless, knowing that like him I had been a soldier in the First War, suddenly began a moment-to-moment account of an attack at dawn. I listened in amazement. Not the smallest detail—the very dew on such grass as remained in No Man's Land—was missing from it. A novelist of genius could not have matched it under a month's hard work. And this man was not working at it; he went on and on effortlessly and with never a halt; and after the first minute or two he spoke in a curious sing-song manner, markedly different from his manner in our preliminary chit-chat; so perhaps he had somehow hypnotized himself.

What was certain was that behind this effortless total recall was a sudden access to the brain's lifelong recording system. But was the brain opening its own files or had the mind suddenly decided to make full use of the brain's resources? There was not the usual suggestion of memory being hard at work, recapturing a detail here, an impression there. It was as if a time past had been transformed into a time present. A half-hour, years before, did not simply return but opened itself out into an enormous Now. There is something here, I suspect, that will never "be explained in either mechanical or chemical terms." We might ask ourselves why we carry this astounding record of our lives, something that must be more ourselves than we can be at any given moment, nowhere apparently but into the grave and oblivion. I ask again—*Why?*

No doubt it is all too easy for a writer to overestimate the importance of the subject he has chosen. Nevertheless, just because he is so sharply aware of this subject and its influences,

he may notice things that other and perhaps better writers have missed. For example, Lewis Mumford's *The Condition of Man* is a superb performance, but I think it might have been better still if he had brought the idea of Time into his analysis of modern society. Many of the chief features of our society seem to me to have been shaped by this idea, of which that central Fortress is the guardian.

And here I must make a personal point. I am by temperament and conviction a political and social radical. I do not dislike modern society simply because it *is* modern. I have no dreams of a restored hierarchic society in which men of letters will mingle on equal terms (though they never did) with a cultivated aristocracy, in mansions and parks maintained without protest by a dream peasantry. I am to some extent what Americans used to call "a forward-looker." I do not hate, fear, or despise that Fortress and the ideas it represents because they are up-to-date and I know that I am not.

Indeed, as I have suggested earlier, I think they are old-fashioned, like many things that guard a gigantic vested interest, and that even in science the boldest and most open-minded theorists and researchers have left the Fortress behind. Men of this sort do not ask for this kind of security. They do not want the protection of dogmas and anathemas.

The fact remains, however, that the Fortress occupies a position and operates on a certain central level that enable it to exert the maximum possible pressure and influence. Therefore we ought not to be surprised if its idea of Time has done much to mold and shape those features of our society condemned by Lewis Mumford and other social critics.

There is, of course—I have mentioned it more than once—this tremendous emphasis on the future. It has to be the future because there is nowhere else to go. Now there is a reasonable reformer's concern with the future that, if we are willing to plan properly and begin to use our energy and resources intelligently, could bring men more and better food, clothing, housing, and education. But the Fortress emphasis on the future goes much further than this. It is really irrational and magical. It says in effect that the next length of passing time will be different *in quality* from the passing time we know. It is as if that old eternal dream time has been conjured out of the sky and laid along the time-track further down the line, just out of sight. (And there is no difference here between capitalist West and communist East, now glaring at each other across the same dubious common ground.) We are sold this conjuring trick because we love our children and grandchildren, so that we try to believe that sooner or later they will know and enjoy a time quite different in quality from ours.

And if all this seems overstated, try reading those books in which scientists and technologists, especially the latter, describe the future. It is not so much what will happen that is significant; it is the atmosphere, the ambience, in which these things will happen; the magic shows through; the Golden Age will have begun. Nobody is bored, discontented, feeling stale or frustrated, in those supersonic airplanes or spaceships, in those apartment buildings a mile high, in those roofed-in cities. This is the ancient Great Time pulled down into history. It is a confidence trick, a swindle.

This does not mean that the scientists and technologists, with their "happy ever after"

beginning in the year 2000, are deliberate confidence tricksters, swindlers. They are mostly rather innocent schoolboyish types, feeling no sense of frustration; their work is happily absorbing; they feel they are creating something; they enjoy a sense of power. After all, this is their age. So they feel easy and confident in the Fortress. But its dogmas and decrees, though consciously accepted, leave many people outside feeling uneasy and dissatisfied, as if they were being cheated but did not know how and why.

Now one reason for this feeling—though I do not pretend it is the only reason—is that such people are still haunted by older ideas of Time. Something that might once have had breadth and thickness has been reduced to a single thin line. A different quality of experience has been lost along the way. So these scientific and technological rhapsodists of the future, aware of their fellow citizens' unease and sense of frustration and anxious to be of service, offer quantity as a substitute for quality. There will soon be three of everything, 10 times as big. Then life will be different, and we shall not have long to wait.

There is in this, I suggest again, a certain schoolboyish innocence, which men retain when they are excited about their work and their wonderful plans for the future; but there is, too, on a far deeper level (of which they are not aware if only because they do not believe it exists), the same Time-haunting, the same stirring of ancient ideas and vague but enduring myths, that compels them to endow the future with a quality that cannot be found in passing time.

Writers, imaginative men, capable of hypnotizing themselves, can be so deeply enthralled and bewitched by the triumph of techniques and machines, that they jump ahead even of these scientists and technologists. One of these—and one of my most charming hosts in the Soviet Union—told me gravely that man himself, the biological creature, had now been so far outstripped by his technical achievements that he would have to make haste to catch up, ridding himself of his physiological deficiencies. As he continued in this vein, I began to see monsters out of science fiction—men-machines or machine-men. And I was reminded of a tele-

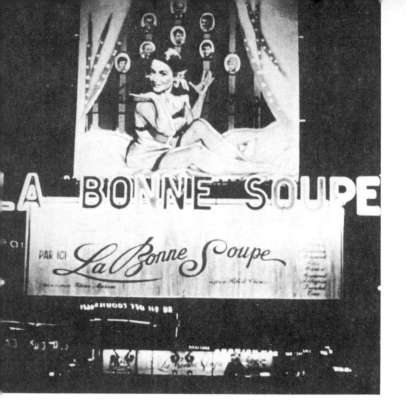

offers us a picture of modern humanity as a restless and ruthless immature male. He is restless and ruthless because the relation between the conqueror and the conquered is rarely satisfying. The conqueror has to push on because he soon finds his victories hollow and disappointing. Nature, nothing if not feminine, seems to take a sly pleasure in outwitting her conquerors. It is the lover and then the husband, not the masterful victor, to whom she yields her treasures. Too many of our conquests are beginning to leave behind sullen wastes of half-occupied territory.

This is not an argument against scientific improvement or bold technical innovations. What is wrong is the attitude of mind that cannot help but think, write, and talk in terms of conquest. This gives us away. Hag-ridden by our idea of Time, the narrowest and worst man has ever had, secretly feeling humiliated, we have to be conquerors, bulldozing our way toward a receding glorious future. The very newspapers that celebrate the latest conquest in their largest print also keep asking us what is the matter with Woman nowadays, when she has so much she never had before. The answer is that she is tired of having to live with neurotic, immature conquerors. But then she is not so driven and harried by this idea of Time. Even now, though perhaps in secret, she still has her own ideas of Time.

By what the Fortress men would instantly call "a coincidence," a term they understand much better than I do, when I broke off writing a few moments ago I went idly across the room to a pile of books, took one out and began turning the pages, not consciously looking for anything but giving myself a break from my desk. The book was William Barrett's *Irrational Man*, and I found myself reading his account of William Faulkner's powerful *The Sound and the Fury*. And I read this:

In the course of the brute random flow of detail that is that last day of his life, Quentin Compson breaks the crystal of his watch. He twists off the two hands

vision program I had seen in which their designers proudly put through their paces the machines that could not only calculate but also walk and talk and compose music.

What was terrifying here was not the machines but these men who were no longer *on our side,* so convinced they were that it was only a matter of time before machines would be much better than men. And a matter of time it really is. For if we really are completely contained within passing time, not able to escape from it with any part of our being, then machines that work faster, are more dependable, and last much longer, may indeed be better than men. But that will only be when we have completely forgotten what men really are, when we have disinherited ourselves perhaps for ever.

If we do not live now, when do we begin? The answer often seems to be "When we have conquered everything." This is very much the Fortress style. Conquest is all. Having conquered land and sea, we are now conquering space, more and more of it. As soon as we have conquered the moon, we must make plans to conquer the planets. I am not denying that there is in all this a fine mixed element of enterprise, ingenuity, and adventure. But the repeated use of these terms *conquer* and *conquest* is worth noting. It

and thereafter, throughout the day, the watch continues to tick loudly but cannot, with its faceless dial, indicate the time. Faulkner could not have hit on a better image to convey the sense of time which permeates the whole book. The normal reckonable sequence of time—one moment after another—has been broken, has disappeared; but as the watch pounds on, time is all the more urgent and real for Quentin Compson. He cannot escape time, he is in it, it is the time of his fate and his decision; and the watch has no hands to reassure him of that normal, calculable progression of minutes and hours in which our ordinary day-to-day life is passed. Time is no longer a reckonable sequence, then, for him, but an inexhaustible inescapable presence. . . . The abolition of clock time does not mean a retreat into the world of the timeless; quite the contrary; the timeless world, the eternal, has disappeared from the horizon of the modern writer as it has from the horizon of modern Existentialists like Sartre and Heidegger. . . .

Now before I had finished copying that quotation from William Barrett's *Irrational Man*, some letters had arrived. They were the last trickles of what had been, some weeks earlier, a torrent of correspondence. I had been interviewed, about this book I was writing on *Man and Time*, on the B.B.C. television program *Monitor*, a late Sunday night program chiefly concerned with the arts. At the end of our talk, the interviewer, Huw Wheldon, had appealed on my behalf to viewers to send me accounts of any experiences they had had that appeared to challenge the conventional and "common-sense" idea of Time.

The response was so immediate and so generous that my secretary and I spent days and days opening letters, then hurriedly glancing through them, as a first step, to sort them out into six categories I had devised. (I shall have to return to these letters later.) This particular letter, which arrived before I had finished the last paragraph, begins as follows:

I wish I could agree with you that the general climate of opinion is now becoming sympathetic towards a serious study of Time. When your fascinating conversation with Huw Wheldon was televised recently I was staying with relations in Wales, and whenever the topic was announced the room emptied rapidly, everyone seeming to find an immediate reason for going to bed. One of my companions (an engineer and technologist) turned back to advise me almost nervously, "For heaven's sake don't sit round listening to that kind of thing. It just makes people morbid—or drives them round the bend." I did listen to what you had to say with extreme interest, and afterwards I wondered—why do normal intelligent people take to their heels at the suggestion that Time be considered and perhaps investigated. Is it because most of us are already aware of strange aberrations

in our own experience, and still react to mystery with primitive fear? Having succeeded in transforming this planet into an animated horror-comic and now turning acquisitive eyes towards other worlds to conquer—haven't we the sense to realise that some of the most intimate and significant fundamentals of our own human nature are still completely unknown, and that it is time something was done about that?...

Now the interview was not based on a script nor even carefully rehearsed, so that I must depend on my memory of hastily improvised answers to questions; but I do not think I went so far as to say that "the general climate of opinion is now becoming sympathetic toward a serious study of Time." What I suggested—or meant to suggest —was that there are signs now in various quarters, not excluding those of the more advanced physicists and mathematicians, that what I have called here the Fortress idea of Time is not good enough. As I remarked earlier, there may be bombardments, coming from different directions of this great central citadel of science, technology, positivism. But there is no formidable army marching against it yet; the best we can say is that some of the territory it overlooks and claims to rule shows us various skirmishing brigades and groups of guerillas, of whom I am proud to be one.

What this letter makes clear—and I could support it with hundreds of others—is the way in which the Fortress exerts its power and uses its influence. It is a new and even narrower Church, stiff and heavy with dogma, and probably already encouraging a quiet scholastic Inquisition. It is determined not to change its mind about anything. However inadequate its idea of Time might be, no matter how the evidence against it begins to pile up, notwithstanding all the harm it may be doing (even an engineer ought to have noticed that our world is hardly in a state of radiant mental health), that idea must not be challenged. "For heaven's sake don't sit round listening to that kind of thing. It just makes people morbid—or drives them round the bend."

Now it is true that more and more people might be called "morbid," and that more and more have been driven "round the bend." We have Fortress statistics to prove this. But we have

no statistics to prove that more and more gigantic communal investigations into the nature of Time are driving people out of their wits. Such investigations are on the most minute scale, and the few people engaged in them seem to me fairly cheerful and responsible citizens. What is making so many people "morbid" and driving them "round the bend" is the kind of world they feel they have to accept now, a world that engineers and technologists do not question and challenge, a world dominated by the worst idea of Time men have ever had. As far as this idea is concerned, it might be better if we went around the bend, escaping at least in imagination—and what is imagination?—from its inexorable conveyor-belt to nothingness. Perhaps truth and reality are always round a bend.

Let me repeat that I must return later to this mass of correspondence; but there is one point, arising out of it, I want to make here. Many of the people who wrote to me confessed that they had been afraid of mentioning to anybody the queer Time experiences they had had—often, though not always, taking the form of precognitive dreams. And I suspect that behind these correspondents were thousands and thousands of other persons who hesitated about, then finally decided against, writing to a stranger on this subject.

Behind them again, making up by far the largest number, I suspect, would be the millions not conscious of ever having had any queer Time experiences because their minds were closed against them. For them the inhibition would be far too strong. Everything they had been taught would be against any recognition of such experiences. And so would the powerful conformity of our age, in which keeping a job may depend upon a person's reputation quite outside any performance of his or her tasks, upon common sense, upon soundness, upon a sane outlook. These are desirable—and crankiness, morbidity, the remotest danger of "going round the bend," are undesirable indeed, not only a bar to promotion but a threat to keeping a job on any but the lowest levels. So a Time experience may be reduced to a mere flicker in the mind, to be shrugged away at once. We do not have to

I KEEP RUNNING BUT I DON'T GET ANYWHERE.

HI THERE HOWARD.

MURRAY'S A GOOD EGG. I SHOULD BE **GLAD** TO SEE HIM GET AHEAD. I WON'T BE BITTER.

HELLO AND GOODBYE — HOWARD.

LUCILLE HAS **LOOKS**. NO WONDER SHE GETS AHEAD. I BET SHE USES HER **BODY**. BUT I WON'T BE BITTER.

WATCH MY SPEED, HOWARD.

IRWIN IS FIVE YEARS YOUNGER THAN ME AND A NO GOOD PUNK!

I KNOW I'D DO BETTER IF NOT FOR THESE STOMACH ACHES.

bother about every silly little coincidence, do we? After all, the people who really know tell us not to worry about such things.

But do they really know? Certainly those of us who are well outside the Fortress may be given to self-deception, a marked human weakness in all ages. But there may be plenty of self-deception well inside the Fortress. And if self-deception has to be risked anyhow, it might be better to risk it to take a broad view than a narrow view. It may be foolish, perhaps even dangerous, to wander too far from the highway and then be the dupe of fantasies; but it may be even more foolish, even more dangerous in the end, to be so determined to keep to the well-tested road that you wear blinkers, see nothing of the surrounding landscape, and find the road itself, all that you can see, more and more wearisome and detestable.

And this may be happening now, when man's last widely accepted idea of Time might well be his very last—as well as his worst. This does not mean he must blow himself off the face of the earth, for he may escape that peril, but only to rid himself, quietly and gradually, of those qualities of mind and heart that make him worth calling a man. The machines will keep chronological time, all he has left himself, better than he can. They will tick away the final glimmers of what was once his glorious imagination, the last faint whispers of his adventure of the spirit.

It could happen. If I believe at heart that it will not, that is because I also believe that we have not come to the end yet of man's ideas of Time, but only of my account to them, too brief, sketchy, obviously inadequate, but meaningful to us in our present situation, as we hesitate at the crossroads.

A strip cartoon by America's Jules Feiffer illustrates the "rat race"— modern man's highly competitive pursuit of material success. Our preoccupation with success and progress may be partly explained by our limited view of time—a view that has reduced many people's existence to a desperate struggle to make the most of their share of time.

Part Three   Examples and Speculations

8   The Letters

At his home in Stratford on Avon, Priestley examines the many hundreds of letters dealing with unusual Time experiences that he received following an appeal on the BBC television program *Monitor* in 1963.

8   The Letters

1.

During the last 80 or 90 years, in one country after another, accounts of precognitive dreams, premonitions, telepathic communications, and the rest have been collected, examined, analyzed, filed, and published on a formidable scale. The documents and the books based on them would together fill—and possibly sink—a canal barge. I propose to use very little of this material. I am not engaged in psychological or psychical research; I am writing a book about Time. And it seems to me that for a personal essay of this kind, letters addressed to me while I am writing it are of more value than letters addressed to other people and then boiled down into evidence. I have already referred, on page 187, to the torrent of correspondence that arrived as a result of the appeal, made on my behalf, in *Monitor*, the B.B.C. Television program. It is to these letters that I turn now.

I offer no careful analysis, no exact figures. If without such treatment they cannot be accepted as evidence, then we shall have to do without evidence. To tell the truth, after they reached 1000 I stopped counting them, because this kind of activity bores me, and my secretary, after helping me with the first great rush, had other things to do. But I have read them all, with the exception of long screeds from obvious madmen, detailed accounts of books I had read myself, and letters from people who thought I was in need of a sermon and not the Time experiences for which I asked. These were added to a pile, not as great as one might have expected, called Odds and Ends.

There were five other categories, labeled alphabetically. Of these the smallest and the least important were *D*—Books Recommended, and *E*—Opinions. Categories *A* and *B* and *C* were divided in the following way:

A, the smallest of all, was for possible examples, outside dreaming, of what I called the *influence of the future on the present*; and I suspect that if I had been allowed on television to explain myself what I had in mind, I might have been sent more examples. But to these we shall return, so I say no more in this place.

The *B* pile consisted of clearly stated and what seemed to me trustworthy accounts of *precognitive dreams*. This pile I read through more than once,

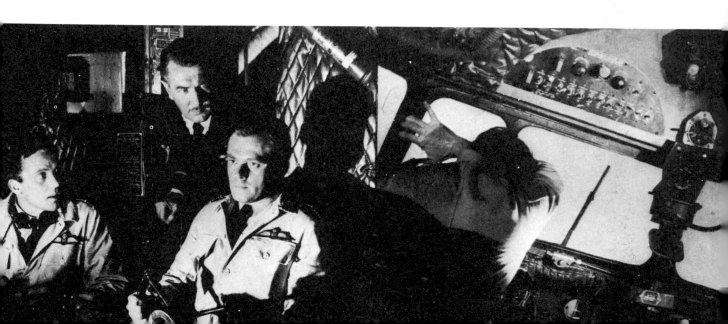

partly to reject some accounts that, on a further reading, did not seem to me good enough, but also to subdivide them into smaller categories.

The *C* pile, easily the largest of all, was made up of precognitive dreams not clearly stated and not sufficiently trustworthy, of premonitions and queer "hunches" that came right, and of various odd little Time experiences not easy to explain but equally not easy to prove. Let me say of this battalion of correspondents, who took the trouble to write to a stranger, that I believe that all but a possible half-platoon were innocent of desiring either to deceive me or to deceive themselves. I regard their letters as evidence of a sort, but for my purpose here not quite good enough. The sheer amount of it, however, is significant.

As I have suggested already (p. 188), many hundreds of these letters support my opinion that there is still popular opposition to the idea of the precognitive dream. Women especially, often mentioning scoffing "down-to-earth" husbands, confessed their eagerness to write to somebody who might believe them. Comparatively few of them were concerned with any theory of Time, even though they might have read—or tried to read—Dunne. They believed in their experience—and in some instances this might go back half a century, my correspondents being of all ages between 18 and the 80s—and so they were glad to pass it on. But only a tiny minority wrote to prove a theory.

And what to me was curious and then significant was that in most instances, when a dream had been told to husbands or (less often) wives or other members of the family or friends or workmates, and this dream had come true, these other people might marvel for a little while but always left it at that. The prevailing notion of Time was not then challenged. Our contemporary idea of ourselves was not questioned. Something odd had happened, that was all; it could not be fitted into the accepted pattern, so it was ignored. Nobody, man or woman, in this great middle range pointed out that if one, just one, precognitive dream could be accepted as something more than a coincidence—bang goes our conventional idea of Time!

It is this great middle range, forming the bulk of my correspondence, that proves my point about the pressure of accepted conventional

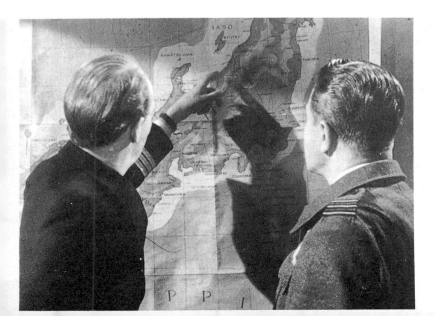

Scenes from the 1955 British film *The Night My Number Came Up*, based on a true experience of Air Marshal Sir Victor Goddard. The day before he was to fly from Shanghai to Tokyo, Goddard overheard a conversation in which he was referred to as dead. When questioned, the speaker described a dream in which he had seen Goddard's plane crash. The man gave details of the plane's three passengers, one a woman, and of the crash. Goddard remembered with relief that the plane was carrying no passengers : But next day three people—one a woman—joined the plane. During the flight, the plane met storms (far left), and was forced to make a crash landing (center left). All survived and were rescued after a search (left).

opinion. Outside it at one extreme were elderly simple people still in the dream-book and fortune-telling world. At the other extreme were the comparatively few men and women who had thought and read about Time—professors or university lecturers, doctors, engineers, senior teachers. But all the others in between were simply determined to relate an experience that they found themselves compelled to believe in, however irrational, fanciful, daft, or morbid it might appear to the people they knew best. They were, we could say, defying the *Zeitgeist*. And it is my belief that if it were not for this pressure of opinion, this strongly inhibiting factor, a vast number of other people would realize they had had such experiences.

It is true, as the representatives of common sense hurry to tell us, that we like to deceive ourselves. But this cuts both ways. Certainly there is self-deceit in favor of appearing unusual, strangely sensitive, "psychic," and the rest. But there is also self-deceit, of a much safer sort, in favor of conformism, sturdy common sense, rationality, and no nonsense. And we need hardly ask ourselves which of these two attitudes is the more fashionable, the easier to adopt, the one more likely to bring good dividends and a sound reputation.

It is, of course, men who are more likely to inhibit themselves for the sake of appearing to be sensible and dependable. So it is not surprising that the proportion of women to men in the *B* and *C* categories is about three to one. (In the *E*—Opinions—the men come loftily into their own.) If pressed I will agree that more women than men wish to appear wonderfully sensitive and intuitive, and may stop being realistic, at least in this matter, in order to deceive themselves and other people. On the other hand, women in general tend to be more realistic and yet in certain matters more open-minded than men, more ready to resist that pressure of opinion. They are less likely to be put into blinkers by ideas.

Moreover—and this is important where precognitive dreams are concerned—as a rule the average woman *notices* far more than the average man; she has a better eye for detail than he has;

she makes a more alert witness. True, where she loves she is more likely to be churned up by emotion than he is, and at such times may be less able than he is to distinguish between truth and fantasy (an important point I shall come to again, later). Finally, it must be said that in spite of their practicality, their realism in ordinary affairs, many women are prejudiced (though perhaps in secret) against a rational and positivist view of life; they want the world to contain inexplicable elements of the irrational and the marvelous; they cannot resist having their fortunes told.

Here I must be frank. There may exist a few superhumanly disinterested intellects, but I believe all the rest of us come down on one side of this fence or the other. In our secret depths, wherever we do our unspoken wishing, either we want life to be tidy, clear, fully understood, contained within definite limits, or we long for it to seem larger, wilder, stranger. Faced with some odd incident, either we wish to cut it down or to build it up.

On this level, below that of philosophies and rational opinions, either we reject or ignore the unknown, the apparently inexplicable, the marvelous and miraculous, or we welcome every sign of them. At one extreme is a narrow intolerant bigotry, snarling at anything outside the accepted world picture, and at the other is an idiotic credulity, the prey of any glib charlatan. At one end the world becomes a prison, at the other a madhouse.

Now we may easily avoid these extremes; but I believe the secret bias to be always there, however much we may pretend to be disinterested and objective; and its influence is always felt when something like precognition is being examined and discussed. We are all at heart on one side or the other, *wanting* it to be true or hoping that it is false. And I began by saying that I must be frank because I am now reminding the reader of my own bias. (I am also asking him to remember that he has one as well.)

Yes, I would rather risk the madhouse than enter the prison. I would rather believe too much—so long as I can do it without bigotry and intolerance—than believe too little. If mistakes are to be made, I prefer to make the mistake of thinking this life too large, complicated, mysterious, wonderful, than to fall into the opposing error and see it smaller and simpler than my own imagination, tidy and all known, tedious. It is better to risk being taken in than to be shut out: Too much credulity can be foolish, too much incredulity may be death.

However, I want to play fair. So here I will put forward the arguments *against* accepting precognitive dreams.

First, we must remember that many people are secretly in love with the irrational and the marvelous. They dearly want something to happen. And if it happens to them, suggesting they are special people, then so much the better. They may not deliberately cheat; they can easily persuade themselves they were prophetic. Dreams and subsequent events may be both touched up in memory so that they match. Many people, for example, have premonitions of disaster, and disaster comes; but on the other hand those same people may have had premonitions when nothing happened afterward, and these they conveniently forget.

A woman may dream that her husband dies, and within a few days or weeks he does die; but then she may have been suppressing from consciousness a deep anxiety about the condition of his heart, so that in the dream this anxiety breaks through and stages his death: There has been no precognition.

Again, some depth psychologists tell us that a strong emotion of a certain kind, experienced in a dream, linking up with a similar emotion, felt in our waking life, can persuade us that the dream and the later event are alike.

Again, the not unfamiliar feeling of having been somewhere before, of apparently playing our part in a whole scene for the second time, the whole *déjà vu* effect, can be accounted for by the brain specialist, who believes in our lobes and not in predictive dreams. We see pictures of places we have never been to; then perhaps years afterward these forgotten pictures pop up to serve as backgrounds to dreams of travel; then when we go to these places we announce we have already visited them in dreams.

And it is possible, as some psychologists have suggested, that we may be so determined to make our dreams prophetic that unconsciously we manipulate our waking life, choosing this, avoiding that, to compel them to come true.

And then there is coincidence.

This is the favorite, of course. To change the image, we might say that scientists and technologists go to their labs and desks through glittering Christmas-tree thickets of coincidences, about which they have no curiosity at all. (They should read Jung's essay on *Synchronicity*, where coincidence at last comes into its own.) But I must continue playing fair, stating the case for the opposition.

We live in the world—as the Chinese used to say—of the 10,000 things. We dream about these 10,000 things. Some people must dream about a combination of these things that later they discover in their waking lives. Let us say they dream about a narrow dark road, a green gate, and beyond it a red front door. Then months later, after looking at innumerable combinations of roads and gates and front doors, they find themselves staring at a narrow dark road, a green gate, a red front door. Precognition? Not at all—simply coincidence. Chance must sooner or later bring one dream, out of thousands, and one waking situation, out of scores of thousands, fairly close together.

But coincidences of this sort must be rare? Certainly, but then so are your so-called precognitive dreams. Why are people not always having them? Simply because coincidences of this sort *are* rare. And when they occur, then fanciful people, who are not satisfied with life as it is and want their wishes to come true, hurry to make the most of them and start blathering about precognition and the problem of Time.

In the preceding paragraphs I have fairly stated the case against precognition, without referring to the underlying belief that it *doesn't* happen because it *couldn't* happen, so that people must be deceiving themselves in the various ways I have suggested. I have yielded the floor to the positivist opposition. Now I am going beyond this opposition, but still attempting to play fair.

Some people reject precognition, together with its challenge to our accepted idea of Time, on quite different grounds. It is not Time but Space, together with its limitation of our senses, that they challenge. It is our extrasensory perception—the famous ESP of all the experiments—that has been at work in these so-called examples of precognition. They are instances, with which Time is not concerned, of telepathy.

Minds are somehow in communication while bodies may be thousands of miles apart. A drowning man or one mortally wounded thinks in anguish of his wife, and she sees him with appalling vividness in a dream or even when she is awake. This extraordinary kind of experience—and examples of it are fairly common—has nothing to do with precognition and our ideas of Time. In other words, we can accept telepathy and ESP without disturbing our notion of unidimensional time, or any belief that our lives are contained by it.

Now Time, not ESP and telepathy, is my subject here, and I have in fact rejected a large number of letters because the experiences they describe, offered as precognitive Time-challengers, could be explained, I felt, in terms of telepathic communication, even when it was less simple and direct than the drowning-man-to-wife instance above. But this does not mean that I agree with those people—and some of them have been devoted researchers—who want to explain *all* apparent examples of precognition in terms of telepathy. According to them, we never see or otherwise experience the future but always take something telepathically from another mind.

It is true that some of these theories insist upon telepathy because without it they find it hard to understand—and here I don't blame them—what exactly it is that is being seen in the future. It has even been suggested that when we see the future we are in telepathic communication *with*

Left, a contemporary drawing of the assassination of Field Marshal Sir Henry Wilson by Irish nationalists in London, June 22, 1922. Ten days before, a friend of Wilson's, Lady Londonderry, had dreamed she had seen him murdered by two men in front of a cottage on a moor. The next day she told two people of her dream. All the details of the actual assassination tallied with the dream—except that it took place in London. (It has been suggested that the murder was plotted in a moorland cottage, which intruded itself into the dream.)

Another well-documented case of a precognitive dream concerns the assassination of Spencer Perceval, the British prime minister, in the House of Commons on May 11, 1812. (Right, a contemporary etching of the murder, and a report from *The Times*.) About nine days earlier a man named John Williams of Cornwall had dreamed of the murder exactly as it occurred. He had the same dream three times in one night and in the next few days talked freely about it. What is especially remarkable is that Williams had no connection with Perceval and did not know him by sight; he knew the man was the prime minister only because someone in his dream had said so.

our own minds in that future. But this last instance should make it plain—and this is something that the other telepathic theories avoid—that we cannot accept telepathy as the sole explanation in the hope of leaving the conventional idea of Time undisturbed. If we can get into telepathic communication with our own minds (or anybody else's) in the future, then the conventional idea of Time must be rejected. And a dislike or fear of this rejection can be found behind most attempts to discredit any notion of precognition. Time must not be disturbed.

I have now given the anti-precognitives a good run, so good that many readers, who believed they were on my side, may now be wavering. Still playing fair, I will now offer two test cases, but they will not come from the letters but from my own experience. Let us see if two dreams of mine, which I believe to have been precognitive, can stand up to the kind of criticism that I have just indicated.

The first dream occurred nearly 60 years ago. In it I saw very clearly, glaring at me out of a vague darkness, the angry face of one of my uncles. I saw very little of this particular uncle in waking life, and that was the only time I ever dreamed about him. He had lightish grey-blue and rather prominent eyes, very suitable for a kind of cold glaring, and, still a child, I was terrified by them in this dream, which was almost a nightmare. I had never seen him even mildly cross in waking life, and did not associate him with displays of bad temper. But now he looked half-mad with rage, as he glared at me across a distance of some yards. He neither spoke nor moved; but then I soon woke up, still feeling terrified.

Many years later, in the middle of the First World War, I was home on leave, and, waiting for the later performance, the "second house," of a music hall to begin, I was having a drink in a rather crowded bar. Suddenly I felt, as one often

does, that somebody was staring at me, and looking down the bar I saw this same uncle, whom I had not seen for some years, and he was glaring at me exactly as he had done in that old dream. For a moment, though I was now grown-up and a soldier, I felt that same childish terror. Then he came across—he had had a few drinks—and began to reproach me angrily about something that was actually no fault of mine. I did not tell him that he had begun his half-mad glaring years and years before, in a dream.

The second dream belongs to the middle 1920s. I found myself sitting in the front row of a balcony or gallery in some colossal vague theatre that I never took in properly. On what I assumed to be the stage, equally vast and without any definite proscenium arch, was a brilliantly colored and fantastic spectacle, quite motionless, quite unlike anything I had ever seen before. It was an unusually impressive dream, which haunted me for weeks afterward.

Then in the earlier 1930s I paid my first visit to the Grand Canyon, arriving in the early morning when there was a thick mist and nothing to be seen. I sat for some time close to the railing on the South Rim, in front of the hotel there, waiting for the mist to thin out and lift. Suddenly it did, and then I saw, as if I were sitting in the front row of a balcony, that brilliantly colored and fantastic spectacle, quite motionless, that I had seen in my dream theatre. My recognition of it was immediate and complete. My dream of years before had shown me a preview of my first sight of the Grand Canyon.

Both these dreams seem to me definitely precognitive or previsionary. Now let us see how they stand up to the anti-precognitive criticism that I outlined earlier.

The Canyon dream occurred before I had read Dunne, and even when I saw the Canyon I was not yet interested in the Time problem. And the uncle dream and its waking experience happened many years before I knew anything about precognitive dreams. In neither waking event was there merely a vague feeling of having seen something before: There was instant and definite recognition, and in both cases—and this is unusual—there had been a long interval between the dream and event. No suppressed anxieties, no strong emotions, can be found here.

I do not see how I could have manipulated my waking life, even unconsciously, to make these dreams come true. In the Canyon dream I did not know that the apparent stage spectacle *was* the Grand Canyon. It can be argued that photographs of the Grand Canyon are not uncommon, and that I might have seen one or several of them years before I had my dream. I cannot accept this argument. To begin with, as anybody who has seen the Grand Canyon knows, color is essential to any picture of it, and the kind of color photography familiar enough these days hardly existed in the earlier 1920s; and if I had seen a painting of the Grand Canyon I would certainly not have forgotten it. Then again, there were in both the dream spectacle and the actual sight of the Grand Canyon a shared richness, depth, intricacy, and magnificence, that make any comparisons with vague memories of old photographs seem ridiculous.

So we are left with coincidence. And this is roughly how the argument will run. It is not surprising that a child should have a rather frightening dream about some older relative glaring at him. Nor is it very astonishing that a young man should encounter an angry uncle in a bar. It is not even a remarkable coincidence that the uncle should look more or less as he did in the dream.

Now this sounds reasonable but it is not. To begin with, I was not always dreaming about glaring older relatives, and indeed this dream is the only one I can remember. Then on no other occasion did I encounter an angry uncle (or aunt) in a bar or any other public place. The odds against both these happening—and showing me a face with an identical strange look on it—seem to me very heavy indeed.

The Grand Canyon, Arizona. When Priestley first saw this vast mile-deep valley he at once recognized it as the "brilliantly colored and fantastic spectacle" that he had seen in a dream about 10 years previously.

Again, turning to the other example, I was not always dreaming about sitting in vast vague theatres and seeing some motionless, multi-colored, brilliant, and fantastic spectacle. (The color is important. We seem to create our dreams as we do our films—in black-and-white if that will do; in color if the dream needs color, as this one of mine did.) This dream was so unusual and striking that I easily remembered it. It put me in the front row of some vague theatre balcony because later, when I caught my first glimpse of the Grand Canyon, I was in fact sitting close to a railing, a circumstance I would not have reasonably anticipated.

And I ought to add here that at the time I had this dream, I had no plans for visiting America and had never given a thought to seeing the Grand Canyon. In fact, when I did go to America and crossed to the Pacific Ocean by rail, I saw the Grand Canyon as the result of a last-minute decision to take advantage of a convenient "stop-over" trip there. Later, I paid it several more visits, went down into it, and wrote about it in my *Midnight On The Desert*. It came to have a great fascination for me, and, in my opinion, this explains why I had this precognitive dream.

Earlier, when I stated the case for the other side, I think I dealt quite fairly with the "coincidence" objection to precognition, making it look as reasonable as possible. But now let us look at it from our side. In the first place, "coincidence" is shaken in our faces as if we had never seen or heard of such a thing before. But we know about coincidences; we have had our share of them; they are nothing new in our lives; and we know they are bound to occur.

We begin to wonder and to ask questions when coincidences no longer *look* like coincidences, when too many details of dream and event agree, when the odds against chance go soaring, and when the whole tone and feeling of our double experience seem to be quite different from anything we know in an ordinary encounter with coincidence. There is a point past which coincidences turn into something else, compelling us to demand an explanation, just as there is a point past which scientific detachment can turn into bull-headed prejudice.

The people who will tell me it is all coincidence did not *see* my uncle's face as I did, did not *see* that theatre spectacle in the dream and then that first glimpse of the Grand Canyon. And without knowing themselves the peculiar tone and tang of these experiences and their strange thrills of recognition, they are like tone-deaf men arguing that string quartets have little effect upon us.

And there is something else. Here, from the first, I have confessed to having an emotional bias in favor of precognitive dreams and similar experiences. But when, for example, Professor H. J. Eysenck, in his *Sense and Nonsense*, dismisses precognitive dreams in what seems to me a perfunctory manner, he does not admit to having any emotional bias *against* them. People who adopt his standpoint pretend to write about an impossibly pure scientific detachment. All they ask, they tell us, is that those of us who are foolish enough to believe in such experiences should subject them to "well-controlled studies" and laboratory tests and the like.

But we may be dealing here with a range of experience that simply cannot be controlled and tested, that withers away when it is brought into the field of scientific experiment and proof. (This does not mean that nobody should try harder, as the young women who run the Psychophysical Research Unit at Oxford are trying.) And it does not surprise me that experimental psychologists —some of them attempting to deal with the psyche as if it were a lump of sodium—do not have precognitive dreams: Their minds are made up against them.

2.

Later sections of this chapter will be devoted to dreams, so here in Section 2 let us take a rest from them. Now the televiewers were asked on my behalf for not only precognitive dreams but also examples of the influence of the future on the present. Very few of these were sent to me. There are probably several reasons for this scarcity. A large number of the televiewers may not have understood what I had in mind. Those who did may never have thought of their own experience in terms of the future influencing the present. The idea may have been foreign to many of the viewers, and I may have been at fault in not having it carefully explained to them. In order not to be at fault a second time, I must do some explaining here.

Somebody is in a queer state of mind, perhaps behaves oddly, and no reason for this can be discovered at the time. Later—a month, a year, 10 years—the cause of this effect reveals itself. Because of where or what or how I am now, I behaved in such a fashion then. (Dreams do not come into this at all; we are concerned now with our waking selves.) And though I describe this effect in terms of the future influencing the present, it can never be understood in the present that is being influenced by the future; it can be understood only when the effect is well into the past and the future that influenced it is now in the present or the immediate past. It has now to be discovered in retrospect, and this makes it less dramatic and memorable, much harder to trace, than the precognitive dream.

That then is the future-influencing-present effect; and, to save trouble and be in the fashion, we will call it FIP. We hear or read very little about it just because it has to be discovered in retrospect and easily goes unnoticed by the memory, and very few people are FIP conscious. Moreover, it is apt to work for intimate relationships that most people prefer not to discuss. Here is an example, from the private lives of two people I know very well.

Dr A began to receive official reports from Mrs B, who was in charge of one branch of a large department. These were not personal letters signed by Mrs B, but the usual duplicated official documents. Dr A did not know Mrs B, had never seen her, knew nothing about her except that she had this particular job. Nevertheless, he felt a growing excitement as he received more and more of these communications from Mrs B. This was so obvious that his secretary made some comment on it.

A year later he had met Mrs B and fallen in love with her. They are now most happily married. He believes—and so do I after hearing his story—that he felt this strange excitement because the future relationship communicated it to him; we might say that one part of his mind, not accessible to consciousness except as a queer feeling, already knew that Mrs B was to be tremendously important to him.

Oddly enough—reversing the usual intuitive situation between the sexes—she shared none of these apparently inexplicable feelings, and indeed regarded him with some disfavor until they had met several times. But from his point of view, this seems to me an admirable example of FIP. And I suspect that if a great many people examined their memories, innumerable other examples would come to light.

Now here is one of the letters. The writer is clearly an intelligent man. Toward the end of the war, when he was on active service in the East, he had a breakdown, followed by several relapses while he was at Cambridge University after the war. Then he married and he feels he owes his final recovery to his wife, a woman considerably older than himself and already the mother, by a husband she had divorced, of two children in their teens. But she "was one of the brave and inspired ones and she baled me out firmly"; he had no more breakdowns; they are happily matched, and have a daughter of their own about to enter her teens. Now for a year before he met his wife or knew anything about her, he used to pass the gate of her country cottage on the local bus. And he never did this without feeling that he and that cottage were somehow related. FIP was working.

Here is an example of a different kind. During the First War a girl out walking at night in London found herself looking up at a hospital, quite strange to her, with tears streaming down her cheeks. Some years later she set up house with a woman friend and they remained together for 25 years. This friend was then taken ill and she died in that same hospital at which the girl so many years before had stared through her inexplicable tears.

Coincidence? But why the tears then? It is possible that my correspondent unconsciously steered her ailing friend into that hospital. It is possible that the crying long ago may really have had nothing to do with the building the girl was staring at. But we cannot accept both possibilities; it would be either one or the other; and in my opinion neither is likely. It looked like FIP from inside the experience, and I think my correspondent is right.

Next, a carefully detailed letter, too long to quote, from a man who in his childhood and youth suffered from attacks lasting a day or so, keeping him prostrate with blinding headaches and nausea. Lying in a darkened room, toward the end of each attack he would experience a kind of passage "through a succession of colors, so vivid that they hurt—the reds, blues, greens, and purples merged and wavered," then they would separate and seem to submerge him in the intensity of their glare. At this point, each time, he would feel fully awake and would vomit, and then would sleep long and soundly and awake feeling refreshed and quite well.

Years later, in World War Two, he was with the R.A.F. in Malaya. Japanese fighter planes shot up a convoy he was traveling with in the mountains. He and the other men were ordered to scatter in the surrounding jungle. "As I burst throught the green maze," he tells me, "I saw a small ravine below me, occupied by Australian machine-gunners. A Jap fighter, swooping low, seemed to be following me with personal intent, and I dived into the security of the ravine. In that moment the world exploded into a hell of color. All the jagged splinters of red, blue, green, and vivid purples caught and swamped me and flung me among the gunners."

A bomb in fact had burst just behind him and blown him into the ravine. The next thing he knew he was being violently sick. Later he was taken prisoner. But never again did he have the old attacks, see the bright menacing colors, and then be sick.

The early events became an elaborate fore-shadowing of what happened in Malaya; they were almost like rehearsals of it. Or we can say that the effect of those few minutes in the jungle was so powerful that just as it haunts him now, long after the event, so too it haunted him long before the event. The explosion, so to speak, went in two different Time directions. But in one direction, the unlikely one from the future to the present, the explosion hit him much harder. Now he remembers, no doubt quite vividly, the whirl of colors and his subsequent vomiting; but long before the event, when he had his childish and youthful attacks, he *saw* the colors and then was actually sick.

Moreover, if the attacks always had to end like that, reproducing well in advance what was going to happen in Malaya, then we could say that the attacks themselves, with their bilious and migraine symptoms, were a kind of excuse and build-up for what had to be the ending, the reproduction in advance of the Malayan bomb effect. You have to be sent home from school in 1931, let us say, to suffer headaches and nausea in a darkened room, so that you can experience the effect of a narrow squeak in 1942.

Let us try another letter, still outside the various heaps of precognitive dreams. In July 1958 the writer left her home in the Midlands to stay with a friend in London. During Matins at St. Martins-in-the-Fields, she began to cry and could not stop. Nor could she explain this either to herself or to her friend. Two days later, on the train from Euston to her home, she found herself crying steadily, now with a sense of foreboding. When her husband and her son met her at the station, she knew that in some way her increasing misery had to do with her son. "Three weeks later," she continues, "this bright, loving, and dearly loved boy was taken ill, and within five months he was dead. He was 19."

What follows in the same letter is not the FIP I have in mind, but it is worth quoting. "On the third day of his illness he said to me, 'A dog is going to bark from a long way off.' A few seconds later I caught the first faint bark coming across the fields. Less than a quarter of a minute after-wards, he said 'Something is going to be dropped in the kitchen, and the middle door is going to slam.' Within seconds, my aunt, who was work-ing in the kitchen, dropped a pair of scissors on the tiles, and the middle door slammed. My doctor arrived a little later, and while it was fresh in my memory I related this to him. He said he had not known of this happening before, but plainly my son's brain was working just ahead of time."

Now experiences "just ahead of time" are mentioned in scores of these letters. Somebody between sleeping and waking sees or hears something, then becomes properly attentive, only to discover that the something is not happening, has not happened, but then does happen within the next minute. I have had these experiences myself, nearly always when dozing during the day, often lying on an hotel bed at the end of a longish journey.

What the boy said to his mother, however, suggests quite a different experience. He tells her that a dog is going to bark, that something is going to be dropped in the kitchen, that a certain door will slam. Now if he has already heard these sounds, how does he know that his mother has not heard them yet but *will* hear them? (The fact that these sounds are of no importance in themselves seems to my mind to increase, not decrease, the value of the scene. I will explain this when we return to precognitive dreams.) If he has not heard the sounds but has knowledge that within seconds they *can* be heard, what form does this knowledge take? Or, putting it another way, how is he aware of the gap between his time and his mother's time? He does not say "A dog is barking a long way off," leaving his mother to discover, when she hears the dog, that his time is ahead of hers. No, he tells her in effect that she is soon going to hear a dog barking.

This suggests that he is somehow aware of two times, his own and his mother's. If, to use the doctor's phrase, his brain is working just ahead of time, then his brain is also working *not* ahead of time—otherwise he could not tell his mother not what she is hearing but what *she is going* to hear. He knows in fact that he is, let us say, 15 seconds ahead of his mother, and from his position in his mother's future (for this is a future even if it only amounts to 15 seconds), with knowledge acquired there, he is able, because he is also sharing his mother's present time, to tell her what will shortly happen. I offer no solution here; I am only demonstrating the Time complications in this little scene.

There are other letters I have set aside as possible instances of FIP. A man buys a very charming house, just outside London. One evening, just before the sale is completed, he takes a relative to look at the outside of the house. As they walk away, he turns to look back at it, and then begins shivering from head to foot, something that has never happened to him before. However, he buys the house and moves in. During the six years he is there, his wife has four miscarriages, six close relatives die, and, having been swindled by his business partner, he has to go into bankruptcy.

Again, an elderly woman doctor writes to say that many years ago she was a house surgeon of a hospital, on call at night by telephone. She lived in a flat separate from the main building, and if she hurried it took her about five minutes to reach her wards. I will now quote her own words:

In those days I slept like the dead and was barely sensible when I spoke to the Night Sister if she called me. However, one night I woke to find myself dressing. I was standing by the dressing table putting my things on quite unhurriedly. I looked at my watch, it was 2.30. I put on my frock and then took a coat from the wardrobe. I took a torch from the drawer and moved to the bedroom door. I was completely awake by now but still didn't appreciate that there was anything odd in the situation. *As I turned the door knob* my bedside phone sprang into life. I heard the Night Sister ask me to come to my ward at once. In reply I said "It's Mr. Jones, isn't it?" An astonished Sister met me at the ward door. She was surprised that I was dressed and that I had come over in record time. She was even more surprised that I had mentioned the patient's name. What astonished her most was to find that I was *not* surprised that Mr. Jones was dead, and died suddenly and quite unexpectedly at *2.20*. He was due to go home later that day, following complete recovery from some minor operation—quite unconnected with the "cause of death." And how on earth did I know it was Mr. Jones—when I had forty much more likely customers in my ward?

She goes on to say that she accepted the experience "as some aberration in time, as I had understood it up to then," and now feels that her "own personal time" must have jumped several minutes ahead. Instead of answering the telephone and then dressing, she dressed and then answered the telephone. But if her own personal time was several minutes ahead, why should there have been this reversal of cause and effect? Why in fact couldn't she have heard and then answered the telephone several minutes ahead?

Or must we assume that she could begin dressing in her own "ahead" time but that the telephone could only function in general post office time? Disregarding now this idea of her own personal time jumping ahead, we could at a pinch justify the FIP effect—the telephone about to ring at, say, 2.31, has already made her get up and start dressing at, say 2.27, thus achieving a four-minute FIP.

This could happen but, remembering Mr. Jones, I doubt if it did; and reluctant though I am to lose a collaborator to the ESP department, I must admit that the best explanation of this whole episode can be found in telepathy. The time jump cannot explain how she knew it was Mr. Jones, because the Night Sister did not tell her—she told the Night Sister. But a telepathic communication, getting her out of bed to start dressing, would begin with Mr. Jones. It may have come from a dying Mr. Jones, a dead Mr. Jones, or the Night Sister; I don't know and I haven't to know; I am not trying to work in the ESP department.

Another example, the last, not striking perhaps but with a curious little twist in it. A man writes to tell me about a precognitive dream he had in his youth, about discovering a blackbird's

nest with three eggs in it, and how, on a walk the next afternoon, something about the lane he was in reminded him of his dream, and then a few yards further on he found the nest and the three eggs. He goes on to say that this, like other dreams he could remember, was trivial, and that no dream gave him warning the previous year when, while touring in Italy, he and his wife received the news that their only son had been dreadfully injured in a road accident, from which at the time of writing this letter the son had not yet recovered.

But now comes the curious little twist. This man and his wife stay on camping sites when they are abroad, and as they are never sure where they will be, they never bother about forwarding addresses. On three tours they had never once sent an address home. Now on this fourth tour, arriving at a site in Florence, for the first time they sent a cable to say they would be at this address until the following Friday. Late on Thursday night they had the news of their son's accident. No warning dream certainly; but what was it that made them break their rule and for the first time send their address?

I shall say again what I said at the beginning of this section on FIP—the future influencing the present. It cannot be understood when it is happening, only after (and often long after) it has happened. Take the example I like best, that of my friends Dr A and Mrs B. If he had not had an exceptionally good memory, especially of his own past states of mind, I would never have discovered this example of FIP. It can only show itself retrospectively. And the present that is being influenced by the future may have nothing in it as dramatic and memorable as a vivid and perhaps sinister precognitive dream; it will probably offer nothing but odd and unaccountable

states of mind; and these may well be forgotten when the future that helped to create them is now itself the present or has glided into the past.

Bearing in mind this FIP effect—and taking care not to deceive themselves—people who are interested in Time problems should muster and examine their memories, especially, I suggest, those concerned with intimate relationships. And then examples of this effect, the strangest and apparently most irrational of all Time tricks, might soon be multiplied. Finally, though I am not ready at this point to explain how the future *could* influence the present, I am willing to announce quite firmly that I believe it does.

3.

Now we can go back to the *B* and *C* piles of letters already described. I have been poking rather gingerly at the *C* pile, which must now add up to between 600 and 700 letters. (It is to this pile that most of the letters still trickling in are added because that is where they belong.) I have also been going through a small book, published in 1937, called *Some Cases of Prediction* by Dame Edith Lyttleton. I do not want to discuss this book, which after all is only one of many, but it does suggest a comparison that

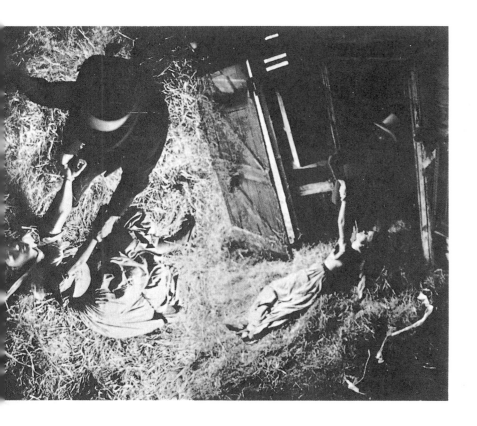

The true incident popularly known as the Red Barn Mystery involves a dream that in a way predicted the future, but that also proved to be an exact reconstruction of events —quite unknown to the dreamer— that had happened many months before the dream occurred. In 1827, Maria Marten of Suffolk, England, ran away with a farmer, William Corder, who betrayed her and murdered her. He buried her body under the floor of the red barn on his mother's farm. (Above left, a contemporary engraving of the barn.) Corder corresponded with the girl's parents as if she were alive, but over the months the parents became suspicious. Then, on three successive nights, the mother dreamed that her daughter had been murdered and buried in the red barn. (Left, the dream re-enacted.) She insisted that the barn floor be taken up— and a sack was found containing her daughter's body. Corder was arrested, tried, and executed. Like the dream quoted on page 203, the mother's dream—however remarkable—can be explained by clairvoyance, requiring no upset of the conventional idea of Time.

seems to me important. Dame Edith, a member of a well-known family, a public woman, President of the Society for Psychical Research 1933–34, did a broadcast talk on prediction and asked listeners to write to her about their own experiences. Now it seems obvious to me, after going through the cases she quotes (and these include examples rightly dismissed as instances of coincidence or telepathy), that the response to her appeal, made in 1934, was very much smaller and far less interesting than the response to the appeal, not made directly by me but on my behalf, in the *Monitor* television program in 1963.

It may be that the Monitor audience was much larger, though it appears late on Sunday and is not one of the B.B.C.'s big peak-hour programs. On the other hand, a radio talk in 1934 had far less competition to face; its title would select its audience; and Dame Edith was a personage. Is it not likely then that the striking difference between these two responses can be explained by a similar difference between the listeners of 1934 and the viewers of 1963?

During these all but 30 years, I suggest, though the pressure of conventional opinion against any idea of prediction may have remained much the same, more and more people, resisting that pressure, have had experiences they are eager to tell, if not to the grinning or yawning family then to *somebody*. I admit I owe something to the Second World War, which (chiefly because it broke through routine, separated people, churned up emotions) provides the background for so many of the examples of precognition, premonition, telepathy, clairvoyance, sent to me. But even after allowing for that second wartime, the difference is still marked and very significant. During these years, something had begun to stir.

From the huge C pile I have pulled letters at random, to test my original judgment, which might have got blurred after hours and hours of opening and reading letters. But so far, so good. In this category, as I suggested earlier, are first the dreams offered as precognitive that might be explained by telepathy (which I do not deny but must keep clear of Time), by coincidence, or by a faulty memory and innocent self-deception, secretly encouraged by a desire for the marvelous. Then, as I said before, there are all the premonitions, the queer "hunches" that came

Far left, the liner *Titanic* during her trials at Belfast shortly before her maiden voyage in April 1912. One man who had booked to go on this journey was the Hon. J. Cannon Middleton, who had dreamed—on two successive nights before the voyage—that he saw the boat sinking and people struggling in the water. When, a few days later, business reasons suggested he postpone his sailing, he gladly took the chance to do so. At this point—a week before the sailing—he told his wife and several friends of the dream. As is well known, the *Titanic* sank on her maiden voyage with the loss of over 1500 lives. (Left, a scene of the sinking from the 1943 German film *Titanic*.) Coincidence is an unlikely explanation here, for it should be remembered that Middleton had the identical dream two nights in succession.

right, the "seeing around the corner in time" that many people profess to be able to do, the waking glimpse of the future, and all the various odd little Time experiences not easy to explain but equally not easy to prove.

Some of these last are inexplicable tiny jumps in time. Here is one of them:

The only time that time became misplaced for me happened when I was working as a maid in a place called Dunraven Castle in South Wales. There were three of us in the servery just after Saturday luncheon —Hans, the odd-job boy, Renate, the senior maid, and me. The floor of the room was a terra-cotta orange colour. I saw Renate pick up a jug, a white jug, of chocolate sauce. As she turned to hand it to Hans, she dropped it. It smashed, and the pastel-brown sauce formed a very definite pattern on the floor, something like the diagram of an amoeba as shown in a school biology textbook. As I looked the whole scene melted and, like a loop of film, started again. It was terrifying! I remember shouting to Renate, as she picked up the jug, not to touch it, and screaming in horror as I watched the sauce make its predestined shape on that orange floor. I tried to explain to them how I had watched the scene take place a couple of seconds before it had—and, of course, they said that if I hadn't shouted the whole thing *wouldn't* have happened.

Well, had it to happen? Or, by seeing it happen, did she *make* it happen? Alternatively again, if she had deliberately restrained herself when Renate picked up the jug, might it then *not* have happened? And where was it happening before it really had happened?

Questions like this spill out like the chocolate sauce, and we shall often meet them again when we come to look at certain dreams in this and later chapters, and we shall not find them any easier to answer.

A neighbor and friend—a university extra-mural lecturer and also an excellent dramatic critic—came around two or three weeks after I had sorted out most of my letters, with what he called some "quick jottings." They tell a story similar to scores I have relegated to the *C* pile as being unacceptable; the difference in this case is that I happen to be well acquainted with the writer, whom I know to be both intelligent and scrupulously truthful.

After remarking that these experiences "are disturbing enough to have made me hide them away from myself and others for a very long time," he goes on:

Since the age of five I have had intermittently very quick previsions of happening which have subsequently occurred. The odd thing about them is that they have always been accompanied by the picture of the name of the person mainly concerned in the event. For example:

Three weeks before the death in an air-crash of the Duke of Kent during the war, I was playing some ball game in the garden of our house in Wales when I had a sudden vision of an aeroplane at its moment of impact with the ground. Just above the "picture" was written as a kind of headline the words "The Duke of Kent."

About a fortnight before the death in a Comet air-crash I had a "picture" of an aircraft exploding in the air with the words "Chester Wilmott" written as in a headline above the picture. I expect you'll remember him as a well-known war-correspondent.

About two days before his death in a car-crash I saw the name of the film-star Bonar Colleano written above the "picture" of a very violent smash-up.

These are three examples out of about ten in all which I recall. They are all, incidentally, of violent death, and always the name of the person involved appears as a kind of headline. The only person I have ever related these occurrences to is Barbara [his wife], and she can confirm some of them. There is no preliminary to having the "picture," each one had occurred during the day, and as I recall, each one has occurred when I was out of doors. The first two or three I took no notice of at the time of having them, but I remember vividly being brought to consciousness of them as a result of feeling no surprise when the "real" event actually happened. In short, I already knew, and when the news was announced it was as if it was "cold" news. . . .

These notes invite several comments. First, we must not be surprised because these "very quick previsions" took the form they did. As Dunne indicated, newspaper front pages can be part of the future too, and they may be what we sometimes see, not the events they describe. Next, even allowing for the prejudice I have not tried to hide, it seems more reasonable to believe my friend than to reject his "previsions." He would not want to deceive me. Nor can I see that he would want to deceive himself. He holds no set of beliefs that would make him wish for the miraculous; he is not proud of having these experiences, which tend to embarrass him; he was not emotionally involved with the Duke of Kent, Chester Wilmott, or Bonar Colleano; he made no attempt to peer into the future, gazed into no crystal balls or pools of ink, received these flashes almost as impersonally as an instrument might do.

I cannot see that telepathy can explain them; nor can coincidence, unless we are ready to believe that he really had thousands of flashes of this sort, and forgot them all except the few that came true. And I am not ready to believe this because it would mean that he was either idiotic or dishonest.

Finally, two letters that seem to be worth quoting in full. For reasons that will emerge later, I decided to reject the dreams they describe as examples of precognition; but both of them are so curious and fascinating that I feel they deserve a place here. The first dream comes, as well it might, from Ireland:

In my dream I was driving my car along a road near my home, and quite suddenly, out of nowhere, it seemed to me, a little girl about three years of age appeared right in front of the car. I did all I could but found it impossible to avoid hitting her. On getting out, I was told that she was dead. I looked at her as she lay in the road, and felt completely shattered, though I had never had a chance to save her from what seemed to me to be her inevitable fate. I must stress that feeling I had of inevitability.

When I awoke, I realized with horror that I had to drive down that road that very morning, on my way to lunch with my youngest daughter, and I decided to be more than usually careful. On approaching the spot, I looked round most carefully for any sign of children, and there were none in sight, only about five women standing at a bus stop. Relieved beyond words, I glanced down at my speedometer to check, and on lifting my eyes, was completely horrified to see, standing still in the middle of the road, the little girl of my dream, correct in every detail, even to the dark curly hair and the bright blue cardigan she was wearing. I was afraid to use my horn, in case I startled her and precipitated what I felt was going to be a fatal accident, so I slowly brought the car to a halt, just beside her. She never moved, but stood staring at me.

Meanwhile, the women in the bus queue made no sign of interest, and no one tried to get such a young child off the busy road. In fact, they seemed more interested in the fact that I had stopped. Feeling very shaky, I continued on my way, and looking in the mirror, I saw that the child was still standing there, and nobody was bothering about her. By the time I got to my daughter's flat, I was over half an hour late.

When she opened the door, she was looking very worried and upset, and said how glad she was to see me safe and sound.

I asked why she had been so worried, as I have been driving for over thirty years, and she looked at me and said, "I know that, Mummie, but you see, last night I had a terribly vivid dream. In this dream you ran over and killed a lovely little girl, dressed in a bright blue cardigan and with lovely dark curly hair!"

Now I will admit that the professional story-teller in me could not help feeling suspicious. With its neat "twist" or "pay-off" at the end, this account of a dream and what followed it seemed too good to be true. So I asked for some confirmation, and received it—not only from the writer herself but also from her husband, to whom she told the dream in the morning and related the waking events in the evening, and from her daughter, who did have in fact the dream she immediately described to her mother. There was, then, this double dream about the mother running over and killing this same little girl. But it is obvious from the behavior of the spectators along the road that there was no actual child there, only a phantasm out of the double dream.

Telepathy, and not any Time effect, was at work here. Whether the daughter took the tragic little-girl-episode from the mother, or the mother from the daughter, we cannot tell. Some people might argue that the mother's dream was precognitive, based on what her daughter would tell her 12 hours later. But this explanation seems to me far-fetched—and why, in any case, should only the mother catch the dream and not the daughter? What is certain is that this fascinating double dream is well worth the attention of ESP experts and researchers.

The second letter describes a dream—or, rather, a series of dreams—and a waking event that together suggest a perfect brief sketch of a Gothic tale of mystery and terror:

From the earliest possible time I can remember I dreamed over and over again that I was walking up a churchyard path—every detail of the church and the churchyard being quite distinct and vivid. Horses were wandering about aimlessly, the one unreal factor of the dream, and my long hair was clinging round me.

Suddenly, I would feel myself being drawn with great force to a graveside. On reading the grave, I would experience a terrible falling sensation, and awake to find myself in a state of terrible depression.

This dream repeated itself over and over again all through my childhood, never differing in any way.

When I was twelve years old I spent a holiday in the New Forest. Whilst cycling home one day after a swim, I found the very church of my dream, exact in every detail, including Forest ponies outside the gate. My hair was loose and wet. I found the grave too. It looked an ordinary grave until I read the description —*Died April 29th 1934.*

The day on which I was born.

The dream never returned to haunt me again.

Now this letter made the professional story-teller in me even more suspicious. I mistrusted at once that neat "shock twist" at the end, that discovery of the gravestone on which the date of death was also the writer's date of birth. It seemed a good device for a piece of fiction but did not make much sense as an occurrence in real life. Even if we accept reincarnation, which I do not, we cannot believe in a turnover of existence so rapid that while one is ending, another is beginning. Nor could I for one believe that somebody who died on April 29, 1934, might haunt in some mysterious fashion another person who happened to be born on that day. Yet I felt that this letter was too fascinating to be ignored, no matter what it turned out to be.

We will call the writer Miss S. I wrote to her, asking for further and confirmatory details, and these she supplied in a sensible and obliging fashion. The recurring dream of her early child-hood had been frequently described to her parents and friends. (This was confirmed later by her mother.) The actual discovery of the grave was then explained in detail by Miss S, with an exact reference to the churchyard and where the grave could be found there. But why didn't she find out whose grave it was? Her explanation—and she was apologetic about it— was that at the time she was frightened and confused: She was only 12 and she found the grave during a thunderstorm; she never stayed again in that neighborhood, which was not near her home, because her relatives no longer lived there; and for some years afterward, rid

A painting entitled *The Dream*, probably by the Italian artist Dosso Dossi (1479–1541). It shows a sleeping woman surrounded by fantastic creatures and by strange images, such as a distant burning city. This artist's glimpse of the world of the unconscious has captured the quality of vividness that is so often stressed in accounts of precognitive dreams.

now of the dream except in memory, she wanted to have done with the whole frightening and mysterious experience.

She concluded: "Somewhere at the back of my mind I believe it was a youth of 19, but this I cannot swear to." I must add that the manner of her letter, a fairly long one, was calm, reasonable, helpful. I must also add that her manner was the same when later she was interviewed and cross-questioned by a woman researcher, about her own age, and therefore more likely to put her at her ease than I might have done.

The final result of our various inquiries and explorations, which included the photographing of some possible graves, can be stated as follows. The approach to the church, the church itself, the graveyard—all of which she actually saw, outside her dreams, only on this one occasion—were as she described them. But there was no gravestone bearing the date April 29, 1934; and indeed, as the rector put in writing for us, there had been no interments in that churchyard during April or May 1934. Two photographs of possible graves were shown to Miss S, who declared at once they were quite different in type (as well as in date, of course) from the one, and the only one, she saw, which unlike them had very clear black letters and figures on a white upright stone.

It is now reasonably certain that Miss S did not actually see that gravestone with her birth date on it. But she equally certainly thought she did. After a long and searching interview, she was acquitted of any conscious faking—though not entirely of some confusion of memories and associations. So now, what are we to make of her story?

First, her dream is different in kind from most of the others described to me by correspondents. To begin with, "it repeated itself all through childhood, never once differing in any way." This repetition itself seems to me to suggest a not very happy and healthy childhood, one turned away from life. The symbolic content of the recurring dream carries this suggestion further.

Horses, for example, are often symbols of vital energy, and in the churchyard, where she always found herself in the dreams, *horses were wandering about aimlessly*. After that, "being drawn with great force to a graveside" seems almost inevitable. The dream kept returning to make a highly dramatic comment on her life situation. But later, having nearly entered her teens and being healthier and more active (the cycle ride, the swim), she suddenly saw—and during a thunderstorm too—the church and the churchyard she had known in these dreams; and then, profoundly disturbed, living for a moment in a waking dream, projecting figures from her unconscious, she appeared to find a gravestone with the date of her birth on it.

In short, she was being symbolically and dramatically reminded that she had been born into this life and must live it. Then, turned toward life, she was no longer haunted by the dream, was free of the churchyard.

Now if Miss S is right when she insists that she recognized the church at once "exact in every detail," then the recurring dream was in part precognitive. Sooner or later (in a manner explained in the next chapter but one) she would come upon that church and the graveyard. But the emotional force of the recurring dream came from her own life situation at the time of dreaming.

In my opinion the dream was not, so to speak, a foretaste of what she would feel during the future waking event. Instead, her experience in the churchyard was shaped and colored by her recollection of those childhood dreams, which were not, as Miss S appears to think, prefiguring the experience. The appearance of her birth date on the gravestone—an illusion—was the dream's farewell to her.

A close analysis of everything she did and thought and felt, after she recognized the church and then went to look at its graves, might prove my point; but it would be impossible after all these years. I wish I had space here for a further consideration of this fascinating example—if only because I know it could be argued that there was a kind of "mirror effect" between the recurring dream and the final experience, half-waking, half-dream. But instead I must now return to my heaps of letters and all their strange cargo of dreams.

A reconstruction of the churchyard
dream quoted on page 209. The
dream, which haunted the girl
throughout her childhood, was
partly precognitive in that the
church's actual appearance tallied
exactly with the dream church. But
it is probable the girl's waking
experience was also to some extent
shaped by her recollection of the
dream. As soon as she recognized
the church, she was perhaps
overwhelmed by the dream's
emotional power, and for a moment
lived in a waking dream state. Thus
she was able to "see" a gravestone
with her date of birth on it,
although in reality none exists.

9   The Dreams

9 The Dreams

So far I have made only two general observations about these letters. First, that they supported my opinion that there is stiff popular opposition to the idea of precognitive dreams or to any monkeying with Time. Secondly, that the bulk of these letters came from people in a great middle range, between the extremes of unsophisticated dream-book elders and of highly educated persons, who had thought and read about Time. It seems to be now that I can make some further more or less general observations, before considering, as we must, a number of sample dreams. We shall gradually move then from a wide to an increasingly narrow view, from correspondents by the hundred to individual dreams, each of which is worth examining for one good reason or another.

The term most often found in these letters is *vivid*. The precognitive dream, it appears, is almost always a vivid dream. If this goes too far, and we wish to keep the opposition quiet, we can say that many people cannot help being deeply impressed by peculiarly vivid dreams easy to remember. In very few instances are vague dreams, only half-remembered, offered as evidence. (A fair proportion of these letter writers, perhaps one in five of those that offer examples worth considering, have read Dunne and have tried, at one time or another, to follow his advice about recording dreams immediately after waking up). There is a general suggestion—though rarely a definite statement on the subject—that these dreams have a quality of their own, making them vivid and easily remembered. They are quite different from the usual jumble of meaningless incident and misty backgrounds. In many instances the dream has been remembered so sharply that the event it prefigures seems by contrast rather disappointingly drab, as if the dream had somehow drained the color out of it.

One correspondent—and I am not blaming him—refers to "a vivid picture" of the dream or nightmare he had, one night in 1938 when he was a young man of 20. The odd thing here is that he actually saw very little, only felt a great deal. Without any preliminaries, the dream-nightmare began with an explosion, and then he "was aware of a stabbing pain" in his left thigh, right fore-arm, and left temple and eye. He goes on: "I was able to see myself in the dream and in the instant before I awoke I saw the blood as it left my wounds. I awoke, rubbing my left eye."

On May 29, 1940, fighting in the rearguard action of Dunkirk, he was wounded by a grenade thrown into a small country inn. He was wounded in the left thigh, right fore-arm, and, by a splinter, in the region of the left eye, which filled with blood. He went to a mirror to test the sight of this eye, wiping it with a handkerchief while he closed the right eye.

As he looked he "experienced a strange feeling of frightened calm because the image I saw, I had seen that night in 1938 in my own bedroom." In the end, then, it was what he saw that seemed most important.

(Incidentally, I can remember only one clear mirror-image of my face in a dream. This must have been at least 25 years ago, and when I awoke I felt strongly that I had seen myself as I would be sometime in the future. My hair, which was dark and plentiful then, was thin and grey on top as it is now, though my face seemed thinner than it is now. However, there may soon be more wear-and-tear, possibly as I near the end of this book.)

I feel I must find a place here for John Lee, before dreams of a very different sort from his will be discussed. Lee was a butler who was sentenced to be hanged in 1885 for the murder of his employer, an old lady. On the night before he was to be hanged, in Exeter Prison, Lee had a dream in which he was led from his cell down through the Reception Basement to the gallows just outside, then placed on the "drop" (the trapdoor), and kept waiting there because the machinery opening the "drop" would not work. Finally, he was taken by another route back to his cell. He related this "very singular and strange dream," as he called it, as soon as he awoke, to the assistant warder and the supervising officer, who had watched him through the night. And these two officers reported the dream to the governor.

Everything happened as in his dream. He could not be hanged and the prison chaplain refused to witness any further abortive attempts. The Home Secretary withdrew the capital sentence and gave him life imprisonment. After serving about 15 years, Lee went to America, not dying until after the outbreak of the Second War.

The dream of course could have been born of a wish not to be hanged, but failures of the "drop" machinery were very rare indeed. Moreover, the route through the prison by which he was returned to his cell, which he described in his dream, was one he did not know in waking life. His prevision was clear and detailed; perhaps because it was so close to the terrible event.

The dreams, no matter how vivid and memorable, are rarely separated from the events that match them by a long time interval. Anything from a day to about a month is the commonest interval. (In this respect my two dreams, already described, were unusual.) Intervals of 10 years or so are so uncommon that I set aside two accounts of dreams just because the dreamers had to wait so long.

A man tells me how, as a boy of seven or eight, he had "a vivid dream" of himself as a young soldier on active service. The scene was a narrow strip of land almost surrounded by water; from a high hill the enemy were firing "black spheres about the size of footballs that glistened in the brilliant sunshine"; and one of these fell near him without exploding, and then two officers appeared, asking him to point out the exact place "as they wanted to take measurements, for a reason I did not understand."

Ten or 11 years afterward, at the age of 19, he was sent in 1915 to Gallipoli, where he found not only the scene of his dreams but also the exact events—the unexploded shell, the two artillery officers who wanted to determine the range of the Turkish gun on Achi Baba—that he remembered from his early boyhood.

The other long-interval example began with a "very vivid dream" when the writer was a schoolgirl just after the First War. She described it to her parents and the art mistress at school. She dreamed she was on a grass-covered terrace above a river, and there were round rustic tables under the trees. "There was," she says, "a quite extraordinary quality about the feeling of light and happiness in the dream." Also, she knew she was wearing a green frock, whereas up to that time she had never been allowed to wear green, because her mother had a superstitious objection to it.

Now 10 or 11 years later, at Easter 1931 when she was convalescent after a serious illness, she went with four friends to spend a day in the forest of St. Germain. "The day," she continues, "had a not unpleasant sensation of waiting-for-something about it but I was still enough of an invalid to be quite willing to drift through the hours. Finally, the one French member of the party suggested our going to the river. So we went—and there was my dream scene—the tables, the trees—my green dress—and above all, the light and slightly floating sensation I had had in the dream."

She insists that the two scenes were not "something like" but really the same; and that, being an artist, she has a very good and trained visual memory. "But it was more than visual," she concludes, "there was the emotional quality too." Quite so; and as "the slightly floating sensation" was part of her convalescent state in 1931, it is all the more remarkable that this sensation, not something seen but something felt, should have been there in the dream. On the other hand, sensations, unless they are very strongly felt, are not easy to compare.

As for the scene itself, I could not help wondering if she might not have come across it, before the dream, in a picture, or a reproduction of one, by the French Impressionists. What came to my mind at once were the Argenteuil-sur-Seine paintings that Monet and Renoir did in the later 1860s. She might have seen one of these as a child, forgotten about it, then recaptured a memory of it in her dream. I put this to my correspondent, who, as she told me, gave my query careful thought before replying.

Up to the time of the dream she had lived in a village near Skipton, in the West Riding of Yorkshire. The only art gallery she had ever visited before the dream was the one in Bradford, which at that period contained no work by the French Impressionists. Her father painted, but his style was "a cross between the Pre-Raphaelites and the later Victorians." Nowadays a visit to any good bookshop might let her see reproductions in color of the Monet and Renoir paintings of Argenteuil, but this was between 40 and 50 years ago, and I never remember seeing any such reproductions myself and I was considerably older than my correspondent. She says herself that she did not "discover" the Impressionists until the 1930s.

She adds finally: "And I still think that the quality of light that I remember so clearly, even after all these years, was such that no painter could capture." In view of this evidence, it seems to me more reasonable to believe than to disbelieve that her dream was genuinely precognitive. And the unusual lapse of time, 10 years or so, between the dream and the waking event, makes this a striking and happy example of a dream coming true.

Deaths, terrible accidents, disasters of one sort and another, loom and threaten in probably rather more than two out of every five of these dreams. Or let us say in about 45 per cent of them. Another 45 per cent of them, at an opposite extreme, are concerned not with matters of life-and-death but with trivialities. Scores of letters begin with an apology for the apparent triviality of the dreams they are about to offer; it is obvious though that their writers, mostly not worrying about any Time problem, cannot escape the feeling that somehow these dreams are important. They are happier, I suspect, when they have got them down on paper and have sent them off to somebody. It is this feeling that compels people to send a stranger an account of some very trivial dream and the tiny incident afterward that confirmed it.

And I thank heaven they do, for within this Time context these dreams and tiny incidents are anything but trivial. They are to my mind better evidence against our familiar concept of Time than most of the dramatic blood-and-thunder previsions of accident, catastrophe, death. They are all the more likely to be true because no strong personal feelings are involved in them. The fact that so often they seem rather silly, hardly worth mentioning, is in their favor.

In themselves they are not dramatic, they are not wonderful, they are not miraculous; they are described because they are *true*. But if they are true, and if they have to be explained in terms of Time, then in the end they may turn out to be dramatic, wonderful, even miraculous.

Let us take a small but clear example from one of these piles of letters. A woman tells three people, with whom she is breakfasting, that she has just dreamed that as they were finishing breakfast a farmer arrived with 33 eggs in a bucket, and that later, as she was standing halfway up the stairs, three more eggs were handed to her. That was her dream. Shortly after breakfast, a farmer arrived and handed her a bucket containing eggs, telling her that there were three dozen. She put them in a basket, paid the farmer, and gave the basket to her husband to take upstairs. (They were packing for a journey.)

A few minutes later, her husband called down to say that her dream was correct because there were only 33 eggs instead of 36 and asked her to go up and count them herself. As she was counting, she was called below, then met on the stairs a woman who said there had been a mistake, three eggs having been taken from the bucket, and now handed her three eggs. Thirty-three and then three eggs in the dream; 33 and then three eggs in the real event. You can call it

À la Grenouillère by Auguste
Renoir depicts a view from a café
on the banks of the Seine. Both
the scene of the precognitive dream
quoted on p. 217 and the actual
scene that later occurred seemed
to resemble the quality of an
Impressionist painting—though
at the time the dreamer had no
familiarity with this particular
school of painting.

coincidence just as you can call it *boojum* or anything else. It is only a matter of a few eggs, very far removed from anything like a world-shattering event. But if you stop clinging to coincidence and try explaining this trumpery affair, you might shatter one kind of world.

The event in your dream may turn up in a way you do not expect. For example, one of my correspondents once dreamed that she and her husband were in what she calls "a motor coach train." They could see the engine through the window. It stopped with a jerk; eventually the driver got out and looked under the wheels; then suddenly the train went on, driverless. It came to a bumpy halt, and the dreamer woke up. In the morning this "vivid dream" came to her mind in every detail, and she recounted it to her family, adding "How absurd! This couldn't happen to a train."

During the evening they listened to the news on the radio; the announcer began one item with "A very curious thing occurred in the South of France"; he then related the events of her dream. So in some curious fashion the dreamer had transformed, in advance, this news item into an apparent personal experience.

Again, the dreamer may have a true prevision of something while creating a fictitious adventure to account for it. The following is a very clear little example of this:

On one occasion I dreamt that a large savage beast leapt up at me putting its paws on my shoulders and thrusting its face within an inch of mine. I thought of this beast as "a tiger" because it was bright orange in color, though its face was rather wolf-like with two pointed erect ears. I was startled in the dream, but not terrified. It was more like a big dog jumping up than a tiger attacking.

On the following day I took my children shopping for a birthday present. While wandering through the toy department, my small daughter suddenly said "Look, Mummy!" and without any warning thrust into my face a large grinning wolf-like head made of sponge rubber, bright orange in color and with pointed ears; a kind of glove puppet without a body, whose jaws open and shut with the movement of the hand. I recognised it at once as the animal in my dream, and of course the experience was the same in that I was startled but not frightened.

This creature was actually purchased by the children, christened *Noser*, and still frequently lurks in my son's bed and gives me an odd feeling when I see that grinning orange face looking up at me from the pillow as if he had materialized out of my dream.

Notice that while, in the dream, she thought "a large savage beast" had come so close to her, she did not feel terrified but only startled, as she did in the toyshop the next morning. There is a queer complication here. While she *thought* it was a large savage beast, she did not behave as if it really were—for then she would have been overwhelmed with terror—but reacted as if somehow she knew it was only a toy creature. Yet she was not conscious of this knowledge in the dream.

Another good example of this type of dream comes from a man in the B.B.C. Engineering Division, who had read Dunne and, following his example, had made notes on what he had dreamed immediately on waking:

Dream experience: I dreamed that (specifically) a sparrow-hawk was perched on my right shoulder; I felt its claws. There was no associated dream incident; this was one of a number of isolated and quite disconnected dream situations on that night.

Waking experience: While I was studying in the lounge of my "digs" (being almost totally absorbed in my reading material) my landlord, who had been clearing out the attic, entered the room with a number of items of wooden refuse which he offered my colleagues as fuel for the lounge fire. One of the items was a stuffed sparrow-hawk mounted on the baseboard of (presumably) an original glass case. I paid little attention, until one of my colleagues, having detached the bird, came quietly up behind me and dug its claws into the shoulder of my jacket with sufficient force to enable it to remain standing on my shoulder. I felt its claws. It was only then that I remembered my dream.

Comments: The probability is, obviously, extremely low; I did not previously know of the existence of this stuffed bird. The interval was short—only two hours elapsed between waking (when I recorded the relevant dream, with other material, in a notebook), and the moment when the waking event took place. I would like to mention that, as Dunne noted, the sensation was of the conviction "I have lived this moment of time before", rather than "I have anticipated this event", and this is also true of other, less spectacular, examples of this effect.

Above, a little girl startles her mother with the toy tiger's head that is described on p. 220. In the dream that anticipated this incident, the mother was "attacked" by a large, orange-colored animal—but one that (since it foreshadowed a harmless toy) had little of the ferocity of a real tiger (right).

I think we can reject coincidence here. Radio engineers are not men who are expecting to find sparrow-hawks perched on their shoulders, no matter whether the birds are alive or stuffed. And the odds against dreaming of a sparrow-hawk on the shoulder and then having a stuffed specimen carefully placed there must be stupendous. It may be asked why this correspondent should dream of the bird being on his shoulder without discovering in his dream how it came to be there, why in fact the dreaming self should be aware in advance of one part of the waking experience and not aware of other parts of it. To this question there is no brief satisfactory answer. But Dunne—though we may disagree with his general theory, as I do—does offer a possible explanation of these odd effects. However, it belongs to another chapter.

I want to return now to the point I made earlier—that in many respects these apparently trivial dreams and the tiny incidents that confirm them, even when the dream and the event in waking life do not exactly match (because, as we have seen, the dreamer may misinterpret the circumstances), offer better evidence against our familiar concept of Time than the more dramatic and striking dreams of catastrophes and death. They are, I say again, more likely to be true because no strong personal feelings are involved in them.

During the period of shock or profound emotional disturbance following the death of somebody much loved, a bereaved person, who dreamed of that death, may unconsciously falsify his or her memory of it to make it appear more definitely precognitive than it actually was. Again, deep and growing anxiety about a loved husband, wife, child, suppressed by consciousness, may invade a dream to show that the worst may happen. Some of the dreams sent to me seem to be better examples of the workings of the Jungian unconscious than of strict precognition.

For instance, one woman relates how her husband had a nightmare: "He was driving a hearse with a coffin in it; it was tearing downhill, out of control, at breakneck pace." When the hearse struck the curb (it is like the nightmare scene in the film *Wild Strawberries*), the coffin fell out and the hat the dreamer was wearing tumbled off, to reveal (though how it is not clear) a blood-soaked bandage around his head. The writer continues: "He rarely dreamed, never had nightmares, and this one upset him so badly he had to tell me of it. Six months later he was dead; he died of a cerebral haemorrhage of the brain. As the dream revealed, he drove himself with damaged head to the grave. . . ." And Jung would have said, I think, that this man's unconscious already knew that he was driving himself too hard; the dream was a warning.

2.

There are surprisingly few examples of dreams that were followed by some action taken in waking life. And several of these few, dealing with skids and similar small accidents in motoring or motorcycling, I felt compelled to discard, as the dreams might very well express in dramatic form a constant anxiety. But here is one that we might consider:

The most vivid and outstanding experience that has ever occurred to me, was one night I woke up in a cold sweat absolutely sure that I had knocked down a small boy with my car. This dream remained with me for two or three days afterwards, but eventually, as all things do, it was put to the back of my mind. It was only a few weeks afterwards, when driving into Manchester, that I had to swerve and brake violently to avoid an accident. I jumped out of the car, and was staggered when I immediately recognised the boy as being the same child as in my dream. . . . I did not hit the boy, but was only inches away from hitting him, which thankfully proved one aspect of the dream wrong.

Three quite different comments can be made on this experience. It could be said that a man who is always driving a car might easily dream of knocking down a child, and that this man deceived himself when he immediately recognized the boy he nearly did knock down as the child in his dream. This is possible, not probable. But if we accept the dream as being precognitive, we can still look at it in two different ways. We can say that the dreamer picked up in advance the incident of nearly hitting the boy, and then, moved by a constant anxiety about such things, carried the incident to a dramatic and perhaps tragic conclusion. So the dream is part prevision and part fiction.

But we can also say that the dream showed a possibility never actualized—that is, so far as its dreadful conclusion is concerned—because, forewarned by the dream, the driver was able to act promptly at the right moment. In short, he was able to *change the future* that had already been revealed to him. These formidable alternatives await us on a later stage of this journey like Scylla and Charybdis.

Another dream is worth quoting because it has an unusual folk-tale atmosphere:

When I was 11 years old, then living in my native Czechoslovachia I had a dream in which I saw a beggar in our house who had been stealing. It was a very small man with a hump.

Some time after this dream (I think it was a few weeks later but I am not quite sure of the time) when I was coming home from school one afternoon, I saw that same man whom I had never seen before in reality, leaving our house hastily, carrying something under his arm. I was struck by the look of this man and remembered at once having seen him in my dream. After having asked my mother very quickly whether something might have been stolen by a beggar during the last few seconds, she discovered that my elder brother's coat was missing from its hanger.

I ran after the little man and with the help of other boys in the village where I lived, we chased him and he dropped the jacket before reaching the woods. . . .

A dream sequence from the film *Wild Strawberries* (1957) directed by Sweden's Ingmar Bergman. While standing in a deserted street, the old man watches the approach of a hearse, and sees it strike a lamp post and tip the coffin into the street. To the dreamer's horror, the lid of the coffin falls open, and reveals the face of the corpse as his own. This film nightmare is a close parallel to the actual dream recounted on p. 222, which in fact shortly preceded the dreamer's death.

This happening made a great impression upon me even as a boy. When I told others about it, they could not understand and I therefore did not mention it again. But it is still very clear in my memory and I can remember the man exactly, the look of his face, little black moustache, black hair, rather large hat, and all the details as if it had happened yesterday . . . and yet this was 44 years ago.

It is like a tale from 444 years ago—the hump-backed beggar hurrying away, the village boys chasing him, the flight into the woods. However, it is a good instance of action taken in real life because of something revealed earlier in a dream.

There are several less picturesque examples. A woman dreamed she was traveling with her dog on top of a bus, the older type of London bus that had a curved stairway. As she was going down to get off the bus, it suddenly started moving, and she was thrown forward, cracking her skull. She remembered the dream on waking but dismissed it because she had no intention of going out. But that morning, finding a large inflamed swelling in her dog's ear, she decided to take him to the animal clinic, going by bus. As she was about to get off, descending the stairs, she remembered the dream and so hung on tightly to the rail. The bus suddenly moved off sharply. She was swung forward and strained her hand, clinging on as she was twisted round, but saved her skull.

Again, a games master at a boys' school dreamed he was walking across the playground, discussing with the captain of the school team the possibility of a certain boy playing in the next match. As they walked through the gateway into the playing field, a small boy ran bang into him. Then—"the actual event took place exactly as in my dream, and as we approached the gate I suddenly realised that this was an exact repetition of the dream and quite subconsciously I braced myself for the impact of the small boy coming through, and he came."

Again, a man with whom I have corresponded on other matters, and know to be reliable, sent me the following:

One day my wife went with our youngest son to lunch out and do some subsequent shopping. Several days later she recalled that she had left her umbrella somewhere, but could not recall where. About three weeks later she had a dream and told it to the boy.

The dream told of a return to the restaurant to enquire whether the umbrella had been left there. The girl in the restaurant replied to the enquiry: "Oh yes, mother took it upstairs and said she would keep it if nobody asked after it as she liked the bird-head handle." In the dream, the girl then fetched the umbrella and handed it over.

On the strength of this dream my wife went with our son and enquired at the restaurant, when the girl *spoke the very words heard in the dream* and handed over the missing umbrella.

Notice that the girl's speech makes it impossible for this to be a dream of a past event—leaving the umbrella in that place and the mother taking it upstairs. The girl's speech is in the future. And the decision to inquire about the umbrella makes the dream precognitive. But if no such action had been taken, what would the dream have been then? We are already in deep waters.

A nice instance of a dream followed by successful action is rather too long and technical to be quoted in full. But here is the gist of it. A medical photographer has to move into a new building and to order a lot of new equipment. Everything arrives except two large developing tanks, for which he searches everywhere, finally concluding they must have been stolen. One night, not long afterward, he dreams that he visits one of the rooms, still not in use, where various odd items are stored. In this room there are some large grey steel cupboards, and in his dream he finds one labeled FJA 39, discovers a key to fit it, opens it—and there are the two missing tanks.

Later on in the day he remembers the dream, goes to the key box and finds one labeled FJA 39, takes it to the room of his dream, opens the appropriate cupboard, sees nothing in there, locks the door and goes away, disappointed. But he cannot escape the feeling that those missing tanks really are in that cupboard. So he tries again, still sees nothing, but this time takes a much closer look. The tanks are there, and he missed seeing them before because the light was poor; they were the same dark grey as the shelves, and had been pushed back.

He adds that he did not store the tanks in these cupboards, had never seen the cupboards before, and knew nothing about the numbers of the locks, the exact FJA 39 he needed coming to him entirely from the dream. He adds also, "Needless to say, no one ever believes me when I tell them the story." To which I add that if he is not to be trusted in this matter, then he should not be employed doing an exacting technical job in a cancer hospital. But then I believe him.

Sir Stephen King-Hall, the well-known writer who was for many years a naval officer, has sent me an account of how he felt compelled to act after a premonition:

One afternoon in 1916 I was officer of the watch in the *Southampton* as we approached Scapa Flow. A mile ahead I saw a small island, and I knew that as we passed it a man would fall overboard. The sea was flat calm. The Commodore, navigator, and other officers senior to me were on the bridge. An officer and a number of men were on the forecastle clearing away the anchor. I thought to myself it must be one of these men who will fall overboard. But what could I do? . . . We got nearer and nearer the island and the feeling grew stronger.

Staking everything on this feeling—and for this reason his experience is of exceptional interest— he gave a number of orders, to put the lifebelts out and muster the sea-boat crew, and so forth. The orders, as would be expected, were soon sharply challenged:

. . . The Commodore said "What the hell do you think you are doing?" We were abreast the island. I had no answer. We were steaming at 20 knots and we passed the little island in a few seconds. Nothing happened!

As I was struggling to say something, the cry went up "Man overboard" from the *Nottingham* (the next ship in the line, 100 yards behind us) then level with the island. 30 seconds later "Man overboard" from the *Birmingham* (the 3rd ship in the line, and then abreast the island). We went full speed astern; our sea-boat was in the water almost at once and we picked up both men. I was then able to explain to a startled bridge why I had behaved as I had done. This is a brief outline of a fully authenticated case of precognition. . . .

And an excellent case too, here gratefully acknowledged. The fact that a junior officer, on active service, took the risk of giving these orders, testifies to the compelling force of this premonition. It might have reached him in the form of a precognitive dream. Arriving when it did, when he was officer of the watch and concentrating his attention on his various duties, it could only force its way into consciousness as a strong feeling, a feeling that refused to be dismissed and even grew stronger. Possibly there was a moment earlier when he relaxed his concentration on the present, so that part of his mind went wandering, as in sleep and dreams, and returned from the future with a premonition about a man falling overboard near the small island.

Actually no man fell overboard from his own ship; it was two men from two other ships; but "Overboard" gave him the essential clue as to what would happen, and the "small island a mile ahead" more or less gave him the time and place; and the action he took was incontestably the right action.

And very similar to such a premonition, I suspect, are the "hunches" that appear to play an important part in the lives of many successful men and women. They are, these lucky people, secret time-jumpers.

What follows now does not come from my own collection but from an article on "Precognition and Intervention" by Dr. Louisa E. Rhine, published in the American *Journal of Parapsychology*:

Many years ago when my son, who is now a man with a baby a year old, was a boy I had a dream early one morning. I thought the children and I had gone camping with some friends. We were camped in such a pretty little glade on the shores of the sound between two hills. It was wooded, and our tents were under the trees. I looked around and thought what a lovely spot it was.

I thought I had some washing to do for the baby, so I went to the creek where it broadened out a little. There was a nice clean gravel spot, so I put the baby and the clothes down. I noticed I had forgotten the soap so I started back to the tent. The baby stood near the creek throwing handfuls of pebbles into the water. I got my soap and came back, and my baby was lying face down in the water. I pulled him out but he was dead. I awakened then, sobbing and crying. What a wave of joy went over me when I realized that I was safe in bed and that he was alive.

A photograph of a line of British cruisers (of the type used in World War I) in the North Sea. In 1916, the British author Sir Stephen King-Hall was sailing as a junior naval officer in a similar convoy. As the ships approached Scapa Flow—the British naval base in the Orkney Islands—King-Hall (who was officer of the watch) gave orders to prepare rescue operations, although no one had fallen overboard. A few minutes later, however, the cry of "man overboard" was heard from each of the other ships—but the two men were quickly rescued due to King-Hall's advance preparations. His action was prompted by a premonition of the event—a premonition so strong that he felt compelled to risk serious consequences to avert the disaster that he had foreseen.

I thought about it and worried for a few days, but nothing happened and I forgot about it.

During that summer some friends asked the children and me to go camping with them. We cruised along the sound until we found a good place for our camp near fresh water. The lovely little glade between the hills had a small creek and big trees to pitch our tents under. While sitting on the beach with one of the other women watching the children play one day, I happened to think I had some washing to do, so I took the baby and went to the tent for the clothes. When I got back to the creek I put down the baby and the clothes, and then I noticed that I had forgotten the soap. I started back for it, and as I did so, the baby picked up a handful of pebbles and threw them in the water. Instantly my dream flashed into my mind. It was like a moving picture. He stood just as he had in my dream—white dress, yellow curls, shining sun. For a moment I almost collapsed. Then I caught him up and went back to the beach and my friends. When I composed myself, I told them about it. They just laughed and I said I imagined it. That is such a simple answer when one cannot give a good explanation. I am not given to imagining wild things.

Now many of the letters I have received describe vivid dreams of places unknown to the dreamer but then afterward visited, when the dreams are remembered at once. What is odd in the example quoted above is that although the dream made such an impression upon the narrator, she did not recognize the general scene when she actually saw it—the sound between the hills, the creek, the pretty little glade. There is no hint of precognition until she sees, for the second time, the baby she is about to leave throwing pebbles into the water. So she cannot intervene, so to speak, until the last moment. The whole stuff of so many precognitive dreams—the prevision of a strange place, subsequently recognized at once—seems of no importance in this experience, which hangs on one dramatic moment.

But if we accept it, as I am disposed to do, then we must also accept one of two things. We must believe that up to the moment when the mother leaves the baby to go and fetch the soap, the dream is showing her the future, but that her return to find the baby drowned is a dramatization of not unusual maternal anxiety: the dream therefore being part-future, part-fiction. Or we must believe that a future containing a dead baby is *changed* by the mother's action, into a future in which the baby does not die and lives to become a father himself: so that of two possibilities, one by deliberate intervention has come to be actualized. This leaves us with a future already existing so that it can be discovered by one part of the mind, and with a future that can be shaped by the exercise of our free will. We cannot have both, we shall be told; it must be either one or the other. Possibly, possibly not. We shall see.

3.

There are so many apparently precognitive dreams of places, most of them following the same pattern, that it would not be worthwhile to quote the letters that describe them. The three that follow, however, break the pattern in one way or another.

In the first, it was something merely read later that provided the dream experience. At the time she had this dream my correspondent was a very junior member of a drama group in West London. At a meeting she attended, the director announced that the next production would be a play by Pirandello, though he was not sure which one it would be. And in those days this girl had not only never seen or read anything by Pirandello but had never heard of him.

The night after the meeting she dreamed she was in Italy, where she had never been, and she found herself in a long narrow room with arches along one side, leading into a rose garden. The floor was elaborately tiled, and at a long table men and women in medieval clothes were eating and drinking. Her partner took her by the hand, raised her from the table, and took her out through one of the arches toward the rose garden. As they passed under the arch, a bat flew in from the garden and she reached up and caught it, holding it in her hand. The dream was "so vivid and made such an impression" on her (no words are more familiar than these, ever since these letters arrived) that she told two or three people about it, the following morning.

Later, she went to the public library and took out a volume of Pirandello's collected plays. To

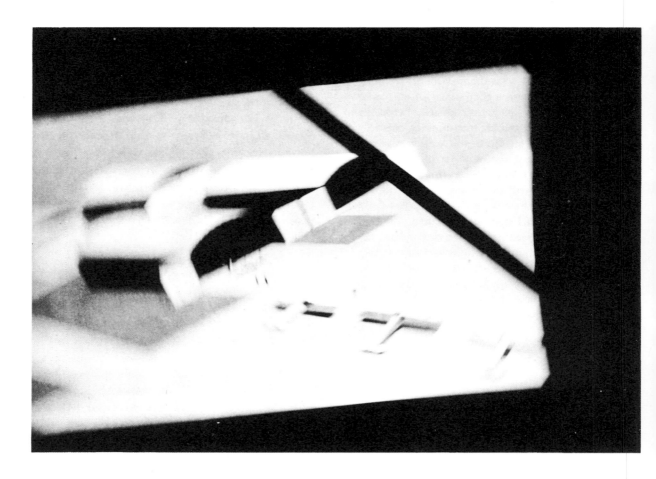

her amazement, "the very first play I started to read has as its setting the exact room of my dream complete with rose garden seen through arches. More amazing still, later in the action—the most improbable stage direction I have ever met—*a bat flies in*." I agree she might have invented a vaguely Italian scene, not unlike Pirandello's setting—but not, I believe, that bat.

The second example is unusual for two reasons. The first is that it seems to have recurred times without number over a period of 10 years or more. (I have many examples of recurring dreams of course—strange to me because I have never had a recurring dream—and they usually take place in the dreamer's early life.) But the dream did not repeat itself exactly for this man. He was always dreaming about the same city, a large seaport, but he would approach it from different directions, sometimes by sea, at other times by rail. He came to have a fairly intimate knowledge of one particular street—"filled with

dives and dubious premises of various kinds." He was acquainted with the people there, noted the death of a certain woman and the arrest of several men engaged in some sort of crime. He spoke the language of these people, and on every dream visit he "was accepted as a local and one of themselves." He began to "know the whole place inside out." For years he tried hard to discover what place this was; then in 1948 he visited Danzig (Gdansk) for the first and only time in his life—"and found beyond doubt that my dream town was Danzig before it was destroyed by shelling in the last war. Since then I have never dreamed about it again."

It is hard to understand why a man should keep on returning in his dreams to Danzig, entering more and more into its life, when he had never been there and was not intellectually or emotionally involved with it in any way. There are two possible explanations. The first is that this queer series of dreams was precognitive, like

A model reconstruction of Drem airfield in Scotland—the scene of a precognitive experience of Sir Victor Goddard (who also was the source of the precognitive dream described on p. 193). While flying in mist and rain over Scotland in 1934, Goddard saw what should have been Drem airfield below him. But instead of the disused hangars among fields that Drem was at the time, the airfield appeared to be in full working order, with blue-overalled mechanics among four yellow aircraft. Four years later, the details of Goddard's experience were exactly fulfilled: The airport was rebuilt, training aircraft were then painted yellow (instead of silver, as formerly), and blue overalls had become standard wear for flight mechanics.

so many other people's, but that what he was receiving in advance was his future experience of reading a novel or seeing a play or film about Danzig. (The writer is not an old man, therefore it is impossible that the novel should have been read, the play or film seen, in the past and then forgotten.) Our dreaming self makes no distinction between first-hand or secondhand experience, between what we shall know directly and what will be communicated to us through fiction or drama, radio or the newspapers. So this explanation is possible.

On the other hand, perhaps the insistent dream experiences came to him through telepathic communication from the mind of somebody who was actually living in Danzig. He was, in these dreams, living somebody else's life; as if lines were "crossed" on the telephone. I have never had recurring dreams of this sort myself, but I have certainly had dreams, entirely realistic and reasonable in their content, that suggested to

me I might be living somebody else's life; but being a professional novelist and dramatist, I have assumed that I must have been using my acquired skills in my dream life. There have been exceptions, however, and I described one of them, years ago, in an autobiographical book, *Rain Upon Godshill*:

One night last year I dreamed myself into some foreign city and though I had no name and did not know what I looked like, I *felt* I was a younger and smaller man, really somebody else, a student or something of that kind; and I crept into a room where there were a number of tiny models of some military or naval invention; and I had just taken one of these from the table when two uniformed officers rushed in, and as I was running out of the opposite doorway one of them fired several times at me, wounding me severely, and as I staggered out into the street I could feel my life ebbing away. I was actually wounded during the war but not in this fashion, and have never in waking existence felt my life fast ebbing away, and I do not believe I could invent that vast throbbing gush of weakness. No doubt most of the dream was my own invention, though I am not given to melodrama of this kind, but I will swear that that swaying progress from the office into the street and the blind weakness that washed over me there were somebody's last moments and that my consciousness had relived them.

One of the most graphic and detailed of the dreams sent to me almost duplicates this experience. It is too long to quote but it describes how the dreamer found himself in a theatre, empty except for a row of condemned men standing on the stage, a firing squad of soldiers, behind the back seats, and their officer, and how the dreamer, after all the men had been shot except one, intervened and was then fatally wounded, and how, clutching at the red theatre carpet he felt his life ebbing away, as I had done in my dream. He awoke with the same impression that I had had, namely, that he had relived somebody's last moments.

We have wandered some way from precognition but can stay away long enough to consider the third dream of a place that breaks the familiar pattern. Here as yet no waking event has arrived to match the dream, but I cannot help feeling that sooner or later it will. The letter begins: "Here is a dream experience which I have had four times during my 31 years: at somewhere around 7, then at 16 and 20 and a few years ago at 28." This dream always takes him into a rough saloon bar, beneath a loft in some wooden shack-type of building. He has seen the surrounding countryside, which is flat, poorly cultivated farmland, showing plenty of rickety fences, tumbledown barns, sagging lengths of barbed wire. Outside the saloon bar is a pile of litter—a large rusty oil drum, broken boxes, tin cans, coils of rope. There is nothing picturesque and romantic either outside or inside this bar. The dreamer does not know where he is—the setting suggests one of the less prosperous regions of the United States, Canada, or Australia.

But each time he has found himself inside that shack he has felt wonderfully content: "I have never experienced such happiness in any life situation so far. It is as if I have arrived at where I wanted to be. I do not speak to anyone, but

they know I am there and that is sufficient. They do not worry about anything or try to be friendly or pleasant as it is unnecessary. They, like myself, belong there. They care about everything without actually caring about anything." He feels strongly that the place exists or has at some time existed, and as he has already visited it at the ages of seven, 16, 20, and 28—incidently, ages very different in their outlooks and ideas of wish-fulfillment—it is not for us to say he has been making it all up. (Why invent exactly the same scene at such different ages?) I should like to think that somewhere this very modest version of a Great Good Place is waiting for him.

4.

Many dreams that may seem nonsensical could be odd glimpses of the future not fully understood. Here are two examples of precognitive dreams to illustrate this. A staff-sergeant serving in a troopship had the usual "very vivid dream,"

in which he found himself lying horizontal in the corner of a smallish room. Two or three women's heads were at his feet. He was not sure of the exact number because he was not looking at them, he was gazing at the ceiling. Nobody spoke; it was all peaceful, satisfying. Then the spell was broken; the door burst open to admit a head, just a head but a man's this time, a noisy joke-making head. The writer observes: "I awoke and genuinely wondered what was reality —the experience I had just had, or the present moment. The 'dream,' as I had to call it, was totally unlike any other dream I had had before. It had a vivid quality of realness about it."

Six weeks later, still on board ship, he had to have an operation for acute appendicitis. The next day, as he was resting comfortably in the top bunk of the cabin and watching the reflections of the water on the ceiling (the ship was at anchor), two nursing sisters came to visit him, then the colleague who made some jokes. It was

the dream all over again, except now he could no longer think of his visitors simply as heads, they were people seen from a top bunk. His dreaming self, seeing them as heads, had thought of them as heads.

In the second example, the lapse of time between dream and waking event was much greater —nine years, an exceptionable gap. In 1928, this correspondent had "a very vivid dream," which he recorded the following morning. It was quite absurd, for in the dream he saw an Indian canoe sailing across the Town Hall Square of his native city. In 1937 he was standing in the Lord Mayor's parlor, in the Town Hall—his father was Lord Mayor that year—and while talking he happened to glance out of the window, which offered him only a partial view of the Square, and—"lo and behold!—serenely floating past was a fully painted Indian war canoe."

Remembering his dream now, he went across to the window to get a better view, then saw a

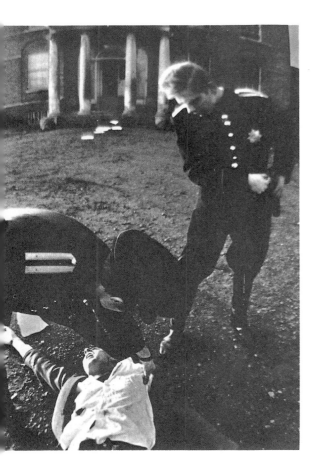

Three photographs recreate J. B. Priestley's strange and probably telepathic dream recounted on p. 229, in which Priestley experienced what seemed to be the dying moments of somebody else—a man who had been pursued from a building and shot several times by uniformed officials.

Within Boundaries, by the modern Australian painter Sali Herman, depicts a shack surrounded by litter in the Australian outback that parallels the setting of the recurrent dream described on p. 230—a dream that may well be a precognition as yet not fulfilled.

number of "props" being carted to one of the local theatres, with the canoe on top. Just as in the other dream the people were accepted simply as heads, so in this dream what was to be later his first glimpse of the canoe, apparently floating past the window, was accepted without question. In dreamland there is no reason why Indian canoes should not sail through the center of a provincial city. In waking life we feel there should be some rational explanation, which this dreamer found there, after a lapse of nine years.

There are several comments worth making here. First, the dream appears to identify itself with a quick glimpse, what the eye will see before the mind offers its explanations, so that heads first seen without bodies remain heads without bodies, a canoe moving across a window must be floating out there. Secondly, it is possible then that what are vaguely remembered as grotesquely absurd incidents in dreams may in fact be actual glimpses of the future taken at their face value and misinterpreted. Of these the canoe in Town Hall Square seems to me a perfect example.

Fortunately, this dreamer recorded his dream. But do all the people who never try to remember their dreams, who dismiss their whole dreaming life, catch similar glimpses of the future that seems grotesquely absurd? We can never know. We must remember, however, that almost all these precognitive dreamers insist upon the peculiar vividness, the queer difference in *quality*, of the dreams they describe. Obviously many of them feel that these dreams-that-come-true are quite removed from ordinary dreams, in a class by themselves.

Now for a further dip into these precognitive dreams. We have seen already how, in a comparatively few instances, a dreamer has been able to intervene, to take decisive action, to prevent the dream prophecy from being completely fulfilled—thus, the baby, at the last moment, is not left alone to play. But sometimes, usually concerning more trivial matters, the dream outwits the waking and alert dreamer. There is an example of this in one of the last letters I have received:

I dreamt one night that I came to the door of the family business in Stroud with one key, and trying to get in discovered that it was still fastened with a second lock. At that moment a rail-car steamed over a nearby railway bridge. On waking I realized that for once this was a dream that I could try to avoid coming true.

That evening I had called a meeting of an unemployed allotment committee on the shop premises at 7 p.m. I had to go home before and collect another member of the committee. The shop door had two locks, one being a Yale lock. I had lost my Yale key and would normally have borrowed my father's key, but he had been away and was not expected back until that evening.

The simple way of avoiding my dream coming true was to arrange with the shop manager to lock the shop at 6.30 p.m. with one key, leaving the Yale lock on the latch. I should then be able to get in by using my one key. This arranged with the utmost care and it was done exactly as requested. On getting home, however, I discovered that my father had returned, and was then driving to Stroud to garage the car and to catch the rail-car home. Immediately I could see that my dream could still come true. Hurriedly, I drove to pick up my fellow member and back to Stroud, explaining on the way my fears. I was soon at the shop door and, jumping out of the car, tried my one key. The door stayed shut. My first thought was to catch my father before he went home by rail. At that moment *the rail-car crossed the nearby bridge.*

My father had gone into the shop to collect post, and finding the Yale lock on the latch had corrected the apparent oversight, and hurried for the rail-car. After thirty years I still puzzle over this dream.

And well he might, though there must be an answer that will explain both this experience of a future that cannot be changed and other experiences in which it has been changed.

Here is another and perhaps trickier one, not from my own collection; it can be found in H. F. Saltmarsh's *Foreknowledge*, a capital little book published about 25 years ago. Mrs. C had a dream in which she found herself being followed by a monkey. As she had that kind of horror of

monkeys which some people feel for certain animals, this dream was a nightmare experience. She spoke of it to her husband and children at breakfast, and he proposed that she should do what she rarely did—they were living in London —that is, take the children for a short walk, so that she might forget this nasty experience. But she had not been out long before she saw, to her horror, "the very monkey" of her dream; and this monkey at once began to follow her.

Now we can see what a neat little box of tricks this is. If she had not described her dream, she would not have gone out at her husband's suggestion, and if she had not gone out, she would not have been followed by the monkey. And if her husband had not wanted her to forget the dream, he would not have suggested the walk on which she met the monkey. The dream, so to speak, was one move ahead of them. They made it come true by trying to get rid of it.

We now arrive at the trickiest of them all, what might be called the double-mirror effect. It turns up in several of the dreams in my letters, but as I have room for only one example quoted in full, I prefer to them the following experience, kindly sent to me by Miss Margaret Eastman, a research officer of the Psychophysical Research Unit at Oxford. I give it in her own words:

In the winter of 1959, when I was still an undergraduate at Oxford, I arranged to go to Cambridge to see the Greek play. A few days before, I dreamed I was in a very ordinary, dingy, suburban room. It was quite unremarkable in every way except for the presence of a white couch. Some people were sitting on the couch, and I was sitting opposite them, talking to them. Suddenly I stopped talking—or rather it seemed as if for a few moments I ceased to be aware of the contents of my own mind though I was still aware of the couch and the room and my presence in the room. Then I started talking again and my mind seemed to start functioning again, and shortly afterwards the dream ended. On thinking over the dream when I woke, I found I could not have said what I was talking about—I was aware simply of having thoughts and expressing them and then of abruptly being locked out of my own mind.

In Cambridge a few days later, I was taken by my companion to visit a friend of his, whom I had never met before and who lived in digs which I had not seen before. The sitting-room was furnished in nondescript brown and beige, except for the couch, which was white. However, I did not recall my dream till later, when I happened to be sitting in an arm-chair talking to the other two who were opposite me on the couch. The realization that I had already dreamed this suddenly struck me and made me pause in the middle of saying something. I then realised that this pause coincided with the part of my dream in which I had been unable to perceive my thoughts. This knowledge agitated me, but feeling that my silence was becoming embarrassing, I forced myself to continue talking.

The double-mirror effect should be obvious. The dream has a blank moment because the waking event it prefigures has a blank moment. But this waking blank comes from remembering the dream and its blank, which itself in turn would not have been there if it were not for the waking blank.

We can put it another way, as indeed Miss Eastman does in her letter to me, and say that in the dream she is already aware of her future memory of having had the dream; but the fact still remains that the dream is so shaped because it reflects what happens in the waking event when she remembers that the dream is so shaped be-cause it reflects—and so on and so on. It is the forced separation of dream and waking event in chronological time that is responsible for this regressive effect.

Finally, the last of the individual dreams, this one from a friendly oldish woman:

Now the dream I would like to tell you of happened 40 years ago, and not to me but to my younger brother. He dreamed of his own funeral and it came about as follows.

At that time my brother was a healthy young man of 24, he played football in a local team. One morning, after chatting and joking over family matters after breakfast, he told me he had had this vivid dream that kept on his mind. He had dreamed of a funeral cortege, of the mourners, the bearers with red and white flowers, even me wearing a wide black hat, and other details, and he was following but no-one took any notice of him. Well, I ribbed him, and we laughed it off as a joke. But the point is, 3 weeks later he got a bad kick on the football field, it turned to peritonitis, he died, and his funeral turnout was exactly as he had dreamed it—the red and white

flowers, his football club colours—and it passed through the same streets, everything the same as in his dream. Nobody believed me when I told them afterwards, they thought the sorrow had upset me, but it happened to be all true, and I have remembered it all my life since.

This young football player was not the first man to have a precognitive dream of his own death. There is the famous case of Abraham Lincoln, who, a few days before his assassination, dreamed that he heard people sobbing, went from room to room without seeing anybody, and finally arrived in the East Room, where his own body was lying in state and he was told, by one of the soldiers standing guard, that the president had been killed by an assassin. (There were several witnesses to his own impressive recital of this strange somber dream.)

Now if these two dreams, reaching beyond the point of death, are taken to be truly precognitive, it follows that the dreaming self and the kind of consciousness it represents must outlast the body by at least several days. There is, however, another possible explanation, one that is good for both the great president and the young English football player.

It is that their dreams were the result of tele-pathic communication, arriving from the future. The football player, I suggest, was sympathetic-ally close to the sister to whom he told the dream; and the funeral he saw was the one she would see in three weeks time or indeed was already seeing in some part of her mind not accessible to ordinary consciousness. (The detail of the "wide black hat" hints at a feminine rather than a masculine glimpse of the occasion.) Either he took the funeral telepathically from her mind as it would be in three weeks time, or the event was already in her mind (suppressed so that she was not consciously aware of it), and it went from her mind to his and shaped and colored his dream.

This means, whichever alternative we prefer, that although telepathy played some part here, so did the Time element. Neither the football player's dream nor Lincoln's can be satisfactorily explained if there is no communication between us apart from our senses and if we are completely contained within passing time.

A seemingly infinite succession of mirror images may be compared to the double-mirror effect of some dreams. In such cases, the dream is a precognition of a later event, but an event that takes place only because of the *memory* of the dream. So, when the two experiences are examined, they cannot be separated to see which caused the other; the dream and the subsequent event seem to reflect endlessly off each other.

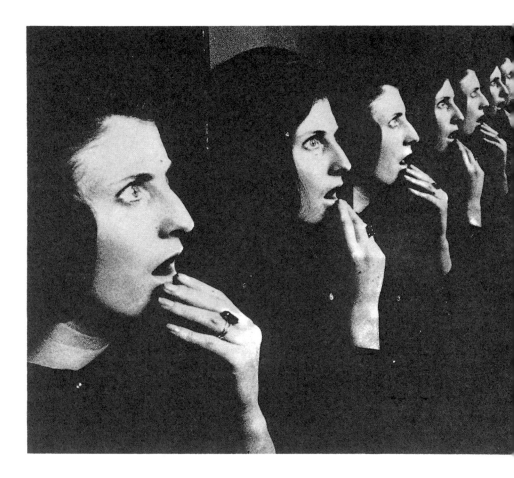

5.

Now we leave individual dreams and what they exemplify. I must continue what I began in the first section of this chapter, a general consideration of all the dreams in these piles of letters. I said there that about 45 per cent of these precognitive dreams referred to deaths, terrible accidents, disasters of one sort and another; that another 45 per cent of them were concerned with trivialities, for which many of the writers apologized in advance. I would add now that rather *more* than 90 per cent of the dreams belong to these groups, the terrible and the trivial: They run to extremes. Sudden deaths and dreadful accidents alternate with previsionary glimpses of seashores or of somebody wearing a blue dress in a black-and-white cafe.

Clearly there is an enormous gap here. A whole wide range of living, between deathbeds and funerals at one end, holidays or shopping expeditions at the other, hardly makes an appearance in all these examples of precognition. And this is so surprising that we must take a closer look.

Among these dreamers, women are in the majority. Now two out of three are married and have children. It would not be unreasonable then to expect a large number of precognitive dreams that showed us a lover, a husband, the arrival of one child or another. The dark young man, the only stranger at the party, might be discovered by the dreamer to be the man she is going to marry. A pregnant woman might have "a clear dream" of the child she is carrying. A quarrel, an encounter with "the other woman," desertion or divorce, all these we might expect to find in women's dreams that involve the future.

But they are not here. Neither are happier sexual experiences. There are no previsions of honeymoons, romantic flights with lovers, silver weddings. And I do not believe this is because women feel shy when writing letters to a stranger. This may be true in some instances, but the im-

Left, in a scene from the American film *The Red Badge of Courage* (1951), Union troops advance to take a position held by Confederate forces. On April 14, 1865, Abraham Lincoln dreamed of a similar victory culminating in the surrender of the Confederate general Joseph Johnston. A few days earlier, Lincoln had had another prophetic dream, in which he saw his coffin lying in state in the White House after the assassination (right).

pression I receive from most of these letters is that if their writers had had precognitive dreams involving such experiences, they would not have withheld them. But they had none to offer me.

As for the men, what is equally surprising about them is not that they avoided any sexual experiences but that, with a very few exceptions, they described no precognitive dreams relating to their work. And the exceptions—though they might be good as examples of precognition—only dealt with trivial incidents. The very things that deeply interest most men—promotion or demotion, a new and difficult responsibility, trying another job, relations with head office or colleagues, all the battles, defeats, or triumphs of a man's working life—are missing from their accounts of precognitive dreams. When we consider how much of a man's time is spent at work, how deeply he is concerned with it on any responsible level, how important its social and financial rewards may be to him, this absence of work dreams is astonishing.

I can only conclude, then, from both the men's and women's dreams, that precognitive dreaming does not offer us any reflection of people's main interests. It fails almost completely to represent this whole wide range of living. There is here in the middle some kind of barrier that is not there at the two extremes of the terrible and the trivial. Precognition exists—and of that I had no doubt long before these piles of letters were here—but only within these curious limitations. Where it might reasonably be supposed to operate most frequently, it hardly makes any appearance at all. It is obviously unresponsive to most of our needs and demands. It refuses to show us the future when we are most anxious that this future should be revealed. (It is true I have received several examples of horse-race winners, but the dreamers were never regular betting folk and, anyhow, coincidence could not be ruled out.) It tells us what we do not particularly want to know—a disaster here, a triviality there—but rarely brings the dream that might help to solve an immediate urgent problem.

I agree that a few precognitive dreams, as we have seen, led to an action being taken to avoid a catastrophe. But these are very few indeed, and appear to be arbitrary. Many people, it is true, have written to say how they have been guided

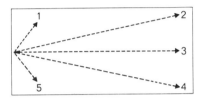
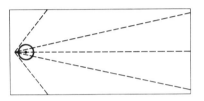

and helped by their premonitions, but these seem often to be so vague and ill-defined that they may be a mixture of foresight and hindsight, and in any event they lack the clear prevision of authentic precognitive dreams.

Such dreams appear to be rare, at least in our civilization. (But odd flashes of the future, mixed with past memories, may appear in many of the muddled nonsensical dreams we usually remember only for a few moments after waking.) While some of my correspondents claim to have had a number of precognitive dreams, many have a clear recollection of one, and only one. This one could arrive at any age past early childhood, and in any circumstances. The whole thing is curiously arbitrary. It is as if the future were a road, or a series of roads, hidden in a fog that occasionally lifts for one dreamer after another. When this happens, the dream is unusually vivid and so is easily remembered. (But the flashes of the future from forgotten dreams may account for the this-has-happened-before feeling mentioned in hundreds of letters.)

Again, the waking experience recalls and matches the dream sharply and unquestionably,

like the right piece in a jigsaw puzzle. But if such dreams and waking experiences are rare, then far rarer still, apparently amounting only to a few exceptions, are the opportunities for taking any action that will in effect change the future. And where is it that we plan and contrive and expect ourselves to act upon the future, feeling that we are helping to create it? Surely it is in that wide middle range of living which hardly ever makes an appearance in these dreams.

What does appear, at each extreme, is something beyond the scope of our planning and contriving and constant intervention. Either it is too big or too small. At one extreme the deaths and disasters, found in so many of these dreams, come crashing through, events so huge and grave that they seem far out of our reach and control. They will happen, and we cannot stop them happening. But this is equally true of the dreams and waking events at the other extreme, where all that is discovered in advance, plucked out of the future, is the sight of a certain sunlit beach or an unusual cafe, some tiny adventure with a toy or a stuffed bird. These too will happen, and we cannot stop them happening, but now it is be-

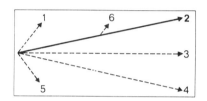

cause they are so small, incidental, unimportant. We cannot arrange our lives to avoid a glove puppet or the claws of a dead sparrow-hawk.

So if the deaths and disasters come crashing through our planning and contriving and shaping of the future, so these trivia—recorded in at least as many dreams, perhaps in more—come creeping in, equally out of our reach and control. We have most foreknowledge of those two borderlands, terrible or trivial, where we plan the least. And where we plan the most, trying hard to shape the future, in that enormous middle range of activities and interests, fore-knowledge seems to be denied us. And what are we to make of this?

Nothing yet. We need still more information before making up our minds. I can, however, ask a few questions. Let us suppose that the dreaming self has some freedom of movement along the fourth dimension, and that it can take, so to speak, a nightbird's eye view of what lies hidden from the waking self (generally, not always) in the future. Why then does this dreaming self so very rarely catch a glimpse of that enormous middle range of our activities and interests?

Here and on the following pages, photographs and diagrams illustrate the interplay of choices and their consequences that makes up the happenings of ordinary life. Such everyday decisions and their results rarely figure in accounts of precognitive dreams, which usually involve either trivial or momentous events. On the middle level between these two extremes, events *may* be foreseen; but here our free choice often intervenes— apparently to *change* the future— so that the events foreshadowed in dreams never in fact take place.

Far left, an aerial view of the junction of five London streets. The dotted lines on the first two diagrams represent the five possible choices confronting the young man in the second picture. In the third picture, the young man chooses the first street (possibility 1, the solid line in the third diagram). In the last picture (above), having chosen the second street (possibility 2), he pauses to speak to a policeman and so misses his friend who is passing by on the right (possibility 6).

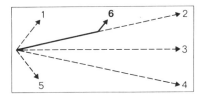

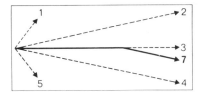

Is it because this dreaming self, for all its wider outlook, discovers here not a determined future but only a confusion of possibilities still to be actualized? And does it observe—to reveal them in so many dreams—the big terrible events and the trivial events just because they are outside this confusion of possibilities, are already welded into the future? And are there then, for this dreaming self (which has a larger outlook than the waking self but one that is still limited) *two futures*—one consisting of determined events, big or little, that can be foreseen in dreams, and the other, seemingly impenetrable to the dreamer, made up of possibilities not yet actualized? And how far are we now from uni-dimensional chronological passing time, the one from which, we have been told so often, we cannot escape?

One final note, to close the chapter. It refers to something as inexplicable as all these precognitive dreams in terms of the conventional idea of Time. It is something that a few of us writers (Thomas Mann, for instance) have noticed, especially when we have been considering other writers, who offer us the best examples because we know more about them than we do about

most other men. It is that some men seem to be aware in a mysterious way how much or little time they have at their disposal, as if deep within their being there was a clock set to tick their years away or a whispering warning calendar.

I have room here for just one example, so I will taken an American novelist with whom I was only slightly acquainted but shared several common friends. This was Thomas Wolfe. A giant of a man, he died at the age of 38; and in his work and his whole way of life he seemed to be wrestling with and raging against a profound conviction that too little time had been allotted to him. Gigantically and frenziedly he worked, filling huge ledgers with millions of words, and he played, talking and eating and drinking and making love enormously, all as if he knew he had, so to speak, to beat the clock.

But what is this clock, marking only so many years, that such men seem to consult in the dark of their being? We do not know. All we do know for certain is that no such clock, no such warnings, can come out of the passing time that we are told is all we have. They belong to a larger idea of Time, like all these dreams that came true.

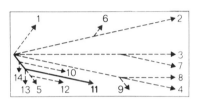

Far left, the young man meets his friend (possibility 6) in the second street (possibility 2). In the second picture, he decides to take the third street (possibility 3) where he meets a girl friend (possibility 7). In the third picture, all three friends converge in the fifth street (having avoided all possible chances of missing one another), consult a newspaper (above right), choose a film, and hail a cab to the cinema.

10　Dunne and Serialism

The great theorist of Time
J. W. Dunne, photographed in
1910 at the controls of a biplane
he designed himself. Dunne's
achievements in aeronautics are
proof of his logical, penetrating
mind. And his scientific
training ensured that he
approached the Time question
with care and detachment.

10 Dunne and Serialism

Until I looked through all the letters sent to me by way of the *Monitor* program, I had not realized how many people had been reading Dunne. Without his examples, and his advice on the immediate recording of dreams, I suspect that at least a third of the best precognitive dreams I have been sent would never have come my way. Although it is not difficult to reject his general theory of Serialism, he remains so far the most important figure in the campaign against the conventional idea of Time. Those of us who are Time-haunted owe him an enormous debt, and I shall try here to pay some portion of it.

Because the Time problem itself is so widely ignored and attempts to solve it are generally shirked, Dunne never became a national figure. During the last war (when he returned to air-craft design) and immediately after it, his health was bad and his circumstances never easy. He was forgotten by everyone except those of us who keep thinking about Time. But we should regard him as one of the world's heroic pioneers.

John William Dunne, the son of General Sir John Hart Dunne, was born in 1875. He served both as a trooper and an infantry officer in the Boer War. He began aeronautical experiments as early as 1900; in 1904 invented the "stable, tailless type of aerofoil which goes by his name"; he built and flew both monoplanes and biplanes of his own type, and finally, in 1906–7, he designed and built the first British military airplane, tested secretly by the War Office in 1907–8. In the early years of the First War, he served as a Brigade musketry-instructor. His very first book was about dry-fly fishing, one of his passions. *An Experiment with Time* was published in 1927, and has been steadily re-printed. His other Time books were *The Serial Universe* (1934), *The New Immortality* (1938), *Nothing Dies* (1940), and a posthumous work, *Intrusions?* He died in 1949. He married the eldest daughter of the 18th Baron Saye and Sele, and had a son and daughter. His clubs, according to an old *Who's Who*, were the United Service and the Flyfishers'.

I mention those clubs because they should help to destroy any possible image of him as a dreamy crank or crackpot. Anybody less like the professional "occultist," the man who knows about "secret ancient wisdom" and "hidden knowledge" and astral bodies and vibrations, it is hard to imagine. As I had been one of the earliest and most enthusiastic reviewers of *An Experiment with Time*, I came to know him, though never intimately; at his own suggestion, he explained his ideas to the cast of my play *Time and the Conways*, a cast that always played well but never better than when they were pretending to understand what he was telling them; and not long before the war I spent the night at the ancient Broughton Castle, when he was staying there with his brother-in-law, and we had our last and longest talk in those fantastic surroundings. He was a slightish man with a good big head; he belonged to the military section of Britain's old upper class, and had its staccato and not highly articulate manner in talk; he looked and behaved like the old regular-officer type crossed with a mathematician and engineer; and, I repeat, he was as far removed from any suggestion of the seer, the sage, the crank and crackpot, as it is possible to imagine.

Though full of ideas and bubbling with enthusiasm, he was no gifted expositor; and in metaphysical debate he was no match for the professional philosophers who sharply disagreed with him. But though he may have made mistakes, as I believe he did, they were honest mistakes: He was a man of intellectual integrity, and as courageous as a thinker and writer as he had been as a man of action.

His greatest achievement was his discovery of what he first called "the displacement in time" in some dreams and his explanation of this displacement, as we shall see later. After this, we Time-questioners owe him most for bringing the

Dunne (right) with actress Jean Forbes Robertson and J. B. Priestley—a photograph taken in London in 1937 when Dunne met the cast of *Time and the Conways* to explain his theories.

whole subject out of various kinds of mystical dusks and magical twilights into the air of ordinary life. Time did not display its tricks only for saints, mystics, holy contemplatives, wizards and sorcerers, and Tibetan abbots 100 years old. Whatever else Dunne may or may not have proved, he did demonstrate that in order to escape from the bondage of chronological time, we did not have to live in some very special and difficult fashion, abandon all familiar pursuits, fast and meditate in the jungle.

To move from one kind of time to another did not involve any mystical exercise or feat of magic; we might not be doing it at will, but we *were* doing it, probably several nights a week. So Dunne argued, after carefully examining his own experience; and I think to argue this successfully, as he did, was itself a great achievement. It changed, quite dramatically, the climate and atmosphere of speculation about Time, taking it out of the philosophers' studies and lecture rooms on the one hand, and, on the other hand, out of the shuttered basements of pseudo-mystics, theosophical magicians, and charlatans.

He may be said to have challenged the conventional positivist idea of Time on its own ground. He did not start from any strong religious prejudices. He had no secret love—as many of us have—of the miraculous. He had not been visited by any mystical revelations. He was not a sentimentally poetic character, outraged by the contemporary world. He was a hard-headed military engineering type, whose hobby was not fantastic speculation and juggling with ideas but fly-fishing. He was, as no doubt they said in the United Service Club, "a sound fellow," even though he did fool around with airplanes in 1906, when sounder fellows knew that these things would never have any real value.

But something happened to him that he could not explain—this "displacement of time"—and just as he had worried away at his aeronautical problems, so now he worried away at this Time business. His final theory may have taken him much too far, but it cannot be denied that he began tackling the problem in a tough realistic spirit, one more genuinely scientific than that of those scientists who had a suspicion that the problem was there but ignored it. Almost all of them ignored Dunne too. Like many another original thinker before him, he was a nuisance.

About his originality, his entirely new approach, the audacity and sweep of his conclusions, there can be no question. He opened a way, and, whatever my reservations may be, I think it is the right way. It is right, to my mind, because he rejects the idea, almost a dogma now, that our lives are completely contained by chronological uni-dimensional time, without becoming other-worldly.

When, for example, we read a book on *Time and Eternity* by Professor W. T. Stace of Princeton, we find this: "The mystic lives in both orders, that of eternity and that of time. He passes from one to the other. This is also true of other men in the degree in which the mystic consciousness is developed in them. But this dual

THE DAILY TE

VOLCANO DISASTER
IN
MARTINIQUE.

TOWN SWEPT AWAY.

AN AVALANCHE OF FLAME.

PROBABLE LOSS OF OVER 40,000 LIVES.

BRITISH STEAMERS BURNT.

One of the most terrible disasters in the annals of the world has befallen the once prosperous town of St. Pierre, the commercial capital of the French island of Martinique, in the West Indies.

ing fuller information about the disaster, and of giving whatever assistance may be necessary.

A STORM OF MUD AND FIRE.

From Our Own Correspondent.
WASHINGTON, Friday Evening.

The following cablegram has just been received by the American Government from Consul Aymé at Guadaloupe:

"At seven o'clock a.m. on the 8th inst. a storm of steam, mud, and fire enveloped the city and roadstead of St. Pierre, destroying every house in the city, and of the community not more than twenty persons escaped with their lives.

"Eighteen vessels were burned and sank with all on board, including four American vessels and a steamer from Quebec named the Roraima.

"The United States Consul and his family are reported to be among the victims. A war vessel has come to Guadaloupe for provisions, and will leave at five o'clock to-morrow."

In an earlier despatch Mr. Aymé said: "I am unable to communicate with Martinique, but am informed that many people are killed there by frequent earthquakes."

The Navy Department has no ship available for relief service at Martinique. As soon as official information is received Mr. Hay will send a message of condolence to France.

Right, a painting of the eruption of Mont Pelée, Martinique, on May 8, 1902. Shortly before this date, Dunne dreamed of just such an eruption (his dream is fully described in *An Experiment with Time*) and was thus not surprised when he later saw a report in the London *Daily Telegraph* (left). The report estimated the number of dead at 40,000; but in his dream Dunne had been convinced that the number was 4000. This led him to the theory that his dream had been a precognitive reading of the newspaper report and that he had simply misread the number by a zero. In fact, later estimates were nothing like either figure.

existence gives rise to confusion of the one order with the other." Now I am not denouncing this book or Professor Stace, an admirable philosophical scholar, if I suggest we need to be rescued from these extreme alternatives—a time too narrowly confining and an eternity too vast and featureless. I do not believe it is true that there are two opposed orders, and that only through "the mystic consciousness" can we escape from the mental handcuffs of the concept of Time we have imposed on ourselves.

We can begin to do it by considering our experience and then comparing its implications with those of that concept. We do not have to dissolve the world or burst it like a bubble. This *either/or* is altogether too narrow, sharp, inflexible. And so it is, to my mind, Dunne's great merit that he rejected this false choice between a world in complete time-bondage or no world at all. Doubting, because it did not fit the facts, what everybody accepted with hardly ever a question, he took a longer and closer look at Time.

2.

In *An Experiment with Time* Dunne carefully describes the dreams he had, beginning as long ago as 1899, that brought him face to face with the Time puzzle. And as they can be found in the book, there is no point in describing them here. They are not very dramatic or striking, and in such respects compare unfavorably with the hundreds of dreams I have been sent. (The severest critics of Dunne's Serialism usually paid tribute to his scrupulous honesty in this account of his dreams.) They had not the atmosphere and emotional tone of fateful previsions. He did not feel he was about to become a prophet.

But gradually—and rejecting on the way any idea of "astral wandering," clairvoyance, telepathy—he came to believe that in some mysterious fashion his dreaming self occasionally caught glimpses of future events, though these might be no more than a glance at newspaper headlines or the reading of a chapter or two in a novel. (It was fortunate, I think, that some of these experiences showed him that he might see in advance not actual events but communications about those events. This was important for the theory he finally worked out.)

By the outbreak of the First War in 1914, he had arrived at this position: "I was suffering, seemingly, from some extraordinary fault in my relation to reality, something so uniquely wrong that it compelled me to perceive, at rare intervals, large blocks of otherwise perfectly normal personal experience displaced from their proper positions in Time. That such things could occur at all was a most interesting piece of knowledge. But, unfortunately, in the circumstances it could be knowledge to only one person—myself."

In 1917, he was in hospital recovering from an operation, and two things happened there of great importance to the development of his theory. One morning, reading a book, he came upon a reference to "one of those combination locks which are released by the twisting of rings

embossed with letters of the alphabet." This vaguely reminded him of something, but after a moment's pause he went on reading. He had to stop, however, and began "determinedly to worry out" what he had associated with that reference to the lock. Then he remembered that he had dreamed the previous night of such a lock: "The chances of coincidence, where two such vague, commonplace events were concerned, needed no pointing out. But I could not remember having seen, heard, or thought of such a lock for a year or more." It was possible then that the merest trivialities, when accompanied by this odd flicker of association, might be traced back to a dream, if the dream could be remembered or, even better, recorded.

The second thing that happened was that two other men in the hospital appeared to have had precognitive dreams, now that the subject was being mentioned. Suppose then that these phenomena were not abnormal but *normal*, and

that dreams—dreams in general, everybody's dreams—were composed of images of past experience blended together with images of future experience?

It should be clear now that Dunne's approach would have to be the opposite of the one I made in the television program. There I asked for experiences—not necessarily precognitive dreams, though obviously they would be in a majority—that appeared to contradict the familiar idea of Time. And what happened I have already described. The letters showered upon me were largely records of exceptional experiences, often vivid previsions that would occur once in a lifetime. But Dunne had to attack the problem from the other side. He had to prove if he could that images of the future played a part—often confused and confusing, therefore difficult to distinguish—in all normal dreaming.

It was not a question of something exceptional and striking but of something that might be found in common experience. This meant that dreams would have to be recalled, and associations with them in waking life (associations often vague and fleeting) would have to be observed. And it would not be easy:

Dreams, moreover, are mostly about trivial things—things which happen every day of one's life. Such a dream, even if it were, in actual fact, related to tomorrow's event, would naturally be attributed to yesterday's similar incident. Then, again, nine-tenths of all dreams are completely forgotten within five seconds of waking, and the few which survive rarely outlast the operation of shaving. Even a dream which has been recalled and mentally noted is generally forgotten by the afternoon. Add to this the before-mentioned partial mental ban upon the requisite association; add to that an unconscious, matter-of-fact assumption of impossibility; and it becomes quite probable that it would be only a very few of the more striking, more detailed, and (possibly) more emotional incidents which would ever be noticed at all. These, moreover, would be attributed to telepathy or to 'spirit messages', or even to anything which, though insane in other respects, could, at least, be expressed in the conventional terms of a single absolute, one-dimensional Time.

And there were two other difficulties. It would be hard to disentangle possible images of the future from images of the past or images symbolizing some present state of mind. For example, you dream you are staying in a farmhouse like one you once knew (image of past); the old kitchen has somehow been transformed into a modern board room (present anxieties); a large confusing group arrive for tea, among them several of your directors (present anxieties) and three colored men arm-in-arm. You may be **about to see these three** (image of the future) but in all this confusion—even though I have added a touch of the unusual—it may not be easy to disengage and record this particular image.

Incidentally, if there should be a lapse of months or a year or two (Dunne did not expect this for his minor effects, which usually followed the dream within a day or so), then if you caught sight of those three colored men arm-in-arm, and the association with the dream was now weak, you would probably be troubled by a vague feeling of having seen them before.

The second difficulty, which Dunne admitted and faced, was that if every detail of dream after dream could be recorded, and if waking events in every detail could be considered in terms of their association with the dreams, coincidence could be expected and could not be denied. But if, as Dunne points out, within one dream episode and the subsequent waking event a possible coincidence is followed by a second, a second by a third, then the odds against the coincidental become astronomical, and chance and random likeness must be rejected.

It must be remembered here that Dunne did not begin with the bias to which I have confessed. He was something of a mathematician, scientist, and technologist himself; his whole background and training discouraged any fancy quasi-mystical notions; he did not start with any suspicion of "absolute, one-dimensional Time," he merely wished to explain, at first for his own satisfaction, certain odd experiences he had had. His attitude was more truly scientific than that of those scientists who hurriedly dismiss any idea that might demand fresh thinking.

Dunne began carefully experimenting himself, and then persuaded various friends and acquaintances—and some of them told him they never did any dreaming—to follow his example. He offers a chapter of advice to those ready to try his experiment, and I will repeat some of it here.

Keep a notebook and pencil, he suggests, under the pillow. Immediately on waking, before you even open your eyes, set yourself to remember the rapidly vanishing dream. If you can remember only a single incident, fix your attention on that incident, and try to remember its details. Make brief notes—and these may have to include incidents suddenly remembered from previous dreams—and later write these out in full, trying to recover and record as many details as possible. "Be specially careful to do this," he adds, "wherever the incident is one which, if it were to happen in real life, would seem unusual; for it is in connection with events of this kind that your evidence is most likely to be obtained."

But there are certain things to be borne in mind. Nights preceding a journey or following a visit to a theatre or cinema are more likely to

Dunne illustrated the infinite regress—a key concept in his idea of Serial Time—by an artist who sets out to paint a picture of the universe. Having painted the landscape before him, he realizes something is missing—himself. So he moves his easel back and paints himself in. But something is still missing—himself painting himself in. So he moves his easel back again—and so on.

produce good results than nights that are part of a dull routine existence: A little excitement helps. Then remember what we discovered in the last chapter—that "the dreaming mind is a master-hand at tacking false interpretations on to anything it perceives. For this reason, the record of the dream should describe as separate facts (a) the actual appearance of what is seen, and (b) the interpretation given to that appearance." (E.g., those two dreams I quoted—the orange-colored wild beast and the Indian canoe.) Remember, too—the clear dream entirely involved with the future being exceptionally rare —that along with a great deal of muddled stuff, adding to the confusion, images of the future may be blended with images of the past.

The comparison of dream detail and waking detail will not be easy. Many dreams will yield nothing, especially in the beginning, but Dunne argues that a closer study of detail often brings surprising results. However, it can be hard going. "To notice," Dunne observes, "that a resemblance between a waking event and a past dream is worth following up, is like trying to read a book while looking out for words which might mean something spelled backwards. The mind cannot keep that up for long." It does not want to keep it up at all:

The waking mind refuses point-blank to accept the association between the dream and the subsequent event. For it, this association is the *wrong way round*, and no sooner does it make itself perceived than it is instantly rejected. The intellectual revolt is automatic and extremely powerful. Even when confronted with the indisputable evidence of the written record, one jumps at any excuse to avoid recognition. One excuse which is nearly always seized is the dissimilarity of the adjacent parts of the scene, or the fact that there are

parts in the 'integration' which do *not* fit the incident; matters which do not . . . affect the fact that there are parts of the scene or integration which *do* fit. . . .

In making this experiment we have to force ourselves to examine our experience in an entirely new way. Dunne suggests that this explains why "this curious feature in the character of temporal experience" has not been recognized before.

He may be right. And certainly this would be a bad time for men and women to be constantly observing that they were seeing, hearing, feeling, what they had already seen, heard, felt, in their dreams. We live in an age that is constantly demanding our sharpest attention. A wrong turn of a wheel or a neglected signal can end our lives hideously. We must attend to things through the the narrowest *now*-point that men have ever known. We are the people whose lives are rushing away in *milliseconds*. No wonder our "intellectual revolt is automatic and extremely powerful," making us refuse to see—unless we hold on grimly to the evidence—that parts of Tuesday's dream and Wednesday's waking experience are identical, and that Time is more complicated than we assumed it to be.

But revolts can work both ways. After putting down, for himself and others, one revolt, the one described above, Dunne now led another—against the accepted idea of Time.

There are some complications in his dream theory, notably the relation between past and future events and their proportionate time lapses, that can be found in later editions of his book and there read with interest; but I do not feel we need go into them here. What is important is what he succeeded in establishing broadly. First, that over and above the exact prevision

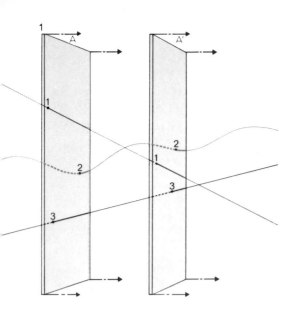

Figure 1 is based on the idea that Time is a fourth dimension, as real as the three dimensions of space. The numbered points in the left-hand plane represent the position of objects or events in space at a specific moment in Time. The right-hand plane exhibits a later moment in Time (the Time dimension is from left to right); the positions of the numbers are different because the positions in space of the objects/events have changed. At each instant the points are stationary; the actual moving is done by the plane—the "present moment" as we observe it—moving in Time. Our memory tells us that the points have moved; but our three-dimensional observation perceives them as stationary. But they are four-dimensional and have "duration" (the crossing lines).

Thus, our three-dimensional observation perceives four-dimensional entities by means of the moving present moment (the moving yellow plane). Now, Dunne contends that, if the present moment moves, there must be a time to time its movement. In figure 2, O represents a specific instant—an observer's "present moment"—and AA′ its movement in Time. To time O's movement requires Time 2—a fifth dimension. Let the movement of Time 2 be up the page. Now, just as the entities had duration in Time 1, so they have duration in Time 2; this is represented by the blue lines (fig. 3). As O moves along AA′ in Time 1, AA′ itself moves in Time 2. Thus O traces a *diagonal,* representing the movement of the present moment in Time 2 (fig. 4).

found in some rare and possibly abnormal dreams, there could be discovered in ordinary normal dreaming, though only as a rule after careful recording and checking, a definite element of prevision or precognition. When we dream we do in fact make use of glimpses of the future, even though not always and not with any particular regularity. Secondly, that our dreaming self therefore cannot be entirely contained within passing time. Thirdly, that there is nothing supernormal and miraculous about this larger temporal freedom of the dreaming self. It is not a privilege enjoyed by a few very strange and special people: It is part of our common human lot. We are not—even though we might prefer to be—the slaves of chronological time. We are, in this respect, more elaborate, more powerful, perhaps nobler creatures than we have lately taken ourselves to be.

A cautious man—as I imagine most aeronautical experts have to be—Dunne did not go as far as I have taken him above until some years after the First War. It must have been in the middle 1920s that he decided he must tackle the whole Time problem. Moreover, he did not propose to use his experiments as the base of operations. They had left him convinced that our conventional view of Time was wrong; therefore he felt he had to reject this view and if possible produce one of his own. This was modesty, not vanity. As he points out, in *An Experiment with Time*, he felt he could not offer his own experiences and his experiments as "scientific evidence." (Note that I am following Dunne's

thought through this one book, using the last edition he was able to correct, instead of dodging about among all his books. This is the most important and the best-known. Later, however, I shall take a quick look at the others.) So although these experiences and experiments, and certain conclusions he had drawn from them, might be said to have taken him some way *inside* the Time problem, he felt it was necessary to return to the outside, beginning again in a new and entirely impersonal fashion.

This was, I believe, a wrong decision. With all respect—and indeed admiration and affectionate gratitude—I must declare my conviction that here he took a wrong and fatal turning. I feel he should never have returned to the outside, never have become impersonal, never have made a philosophical approach to the Time problem. Instead, he ought to have gone burrowing away from the inside, trying to explain as best he could his personal relation to Time. However, with his temperament and training, he might never have been able to do this: Here was a problem that must be tackled as impersonally as some problem in aeronautics.

Yet there was something in him, lying deeper than his professional temperament, that gave him flashes of insight—and these were there from the first, in his attitude toward his dreams—that were far more valuable, and remained so to the end, than all the arguments he mustered. These arguments were soon seen to be vulnerable, and because they could easily be destroyed and then ignored, it was equally easy, in the

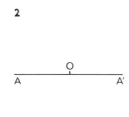

2

3

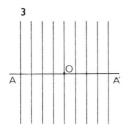

4

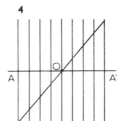

5

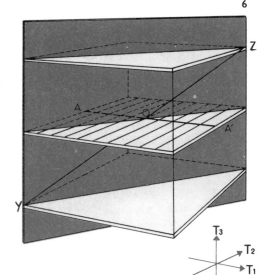

6

The diagonal is the present moment of a four-dimensional observer in a five-dimensional world. This Time 2 observer we can call Observer 2. The movement of Observer 2's present in Time 2 can itself be timed—by Time 3. Tilt figure 4 to show it as a plane (fig. 5) and let the movement of Time 3 be up the page. Then, as the present moment in Time 2 moves up in Time 3 (fig. 6), O traces a diagonal YZ.

This represents the movement of present moments within Time 3, the sixth dimension. But Time 3 can be timed—by Time 4. And so on, in an unending series of times. Of particular interest here is Dunne's contention that Observer 2 can observe the *whole* of Time 1 (AA'), but that normally his attention follows Observer 1's attention in Time 1. Only when Observer 1's attention is relaxed— as when sleeping—can Observer 2's attention wander over Time 1. Thus, Observer 2 could see the future of Observer 1 in a dream.

same critical circles, to shrug away what his flashes of insight had shown him. But though I believe that his metaphysical reasoning was faulty, I also believe that much of what his intuition brought him remains both new and true, and of immeasurable value.

3.

As soon as Dunne reaches Part IV of *An Experiment with Time*, leaving his dream experiences and turning to theoretical exposition, he seems to me to go wrong almost at once. He goes wrong because he insists upon using the concept of Time in two different and contradictory ways. His whole theory of Serialism is erected upon this dubious, shaky foundation. What he does is to spatialize Time and to treat it as movement, both at once. This is having one's cake and eating it.

As I said earlier, I have no objection to a spatial view of Time. Such a view is inevitable, sooner or later, if only as a convenience in argument. What we cannot legitimately do is to dodge between Time lying flat and motionless as an extra dimension of Space and Time in its older role as movement. We must make one approach or the other; we cannot hop between the two. But this is what Dunne did from the first: It is the basis of his main argument.

He begins by telling us that it is never entirely safe to laugh at the metaphysics of the "man-in-

the-street": "Basic ideas which have become enshrined in popular language cannot be wholly foolish or unwarranted. For that sort of canonization must mean, at least, that the notions in question have stood the test of numerous centuries and have been accorded unhesitating acceptance wherever speech has made its way." And this, to my mind, is a suspicious opening. Popular language is a kind of verbal shorthand for quick and easy communication. For example, we know that the earth moves around the sun, and the moon moves around the earth; but in popular language we make no distinction between these movements: The sun rises and sets, the moon rises and sets, and we are still, as we once imagined ourselves to be, in the immovable center of the universe. We all know that this is not true, but we still talk as if it were—simply for our convenience, nothing more.

Now Dunne discovers in this same popular language "two main conceptions—the Time length and the Time motion." He is perfectly right. We do talk as if Time were a length— about a "long" Time and a "short" Time, a "remote past" and a "near future." We also constantly refer to Time as movement, marching on or flowing or hurrying us away. But all such talk is a matter of quick convenience: It no more represents any insight into the truth of things than our similar talk about the sun and the moon.

But Dunne forces it to serve his argument. Because our unthinking talk assumes both a Time length and a Time motion, then we must accept the idea of Time moving along a Time length. And now we are in for it: "For motion in Time must be timeable. If the moving element is everywhere along the Time length at once, it is not moving. But the Time which times that movement is another Time. And the 'passage' of that time must be timeable by a third Time. And so on *ad infinitum*." We are within sight of that infinite regress which was to become the bane of Dunne's Serialism and to which—alas—we shall have to return.

To read Dunne on his dream experiences and experiments and then to read his succeeding chapters on "Temporal Endurance and Flow" and "Serial Time," it is easy to imagine—as I suspect he did—that you are moving from what is doubtful and vague to what is definite and precise. But this would be wrong, at least in this respect: that what appears to be definite and precise does not belong to any acceptable reality. It is only the experiences, the queer previsions, the fleeting premonitions, that are real. Vague and insubstantial though they may appear to be, compared with anything else in the mists and shifting lights of Time theory, they loom up like mountains of iron ore.

We have to remember, no matter where the discussion turns, that dividing Time is like trying to slice up a notion as if it were a Christmas cake. What is real in all this is not Time but our experience—the kind of experience in which we do not know what may happen in five minutes, another kind of experience in which we may get a flash of what will happen next Tuesday, next month, or next year. We are really trying to discover what possibilities we have, what kind of beings we are. Dunne of course was fundamentally aware of this, even though he may have wandered into some verbal traps.

Now, Dunne writes:

. . . We have seen that if Time passes or grows or accumulates or expends itself or does anything whatsoever except stand rigid and changeless before a Time-fixed observer, there must be another Time which times that activity of, or along, the first Time,

and another Time which times that second Time, and so on in an apparent series to infinity. And we might suppose that every philosopher who found himself face to face with this conspicuous, unrelenting vista of Times behind Times would proceed, without a moment's delay, to an exhaustive and systematic examination of the character of the apparent series, in order to ascertain (a) what were the true serial elements in the case, and (b) whether the serialism were or were not the sort of thing that might prove of importance. . . .

Here he is not only grasping the nettle but also poking it into the face of any startled philosopher nearby, for he knows very well that an infinite regress has always been regarded with horror wherever philosophy has been thought and taught. He is enjoying himself for a moment or two in the role of *l'enfant terrible*.

Having had his fun in one short chapter, he now begins a long one in which he is the brisk maths master with a blackboard and a pointer and plenty of diagrams. It is this manner and these diagrams, I fancy, that defeated so many of my women correspondents, who confessed that they read and enjoyed the first half of his *Experiment with Time* but then got lost in the second half. In point of fact the diagrams are clear and fair enough so long as we accept his regressive view of Time. But the verbal trick (if he is simply deceiving us) or trap (if he is deceiving himself) is there, notably when, describing an observer's field of presentation, he goes on to say: "Phenomena in his field seem to move about, alter, and vanish. And these changes appear to 'take Time'"; and then he hammers away at this "taking Time." But these changes *are* Time; or, if we want to be strict about it, they represent the kind of experience from which we abstract our concept of Time. By emphasizing this "taking" of Time, he persuades himself that Time is something like a vast gasoline station, ready to help us to keep moving though always off the scene, somewhere else.

I may believe, as indeed I do, that our blanket term Time may be found to cover two or three different kinds of time; that we may conveniently think of Time being multi-dimensional, using a spatial analogy; but this business of combining Time as a length, a movement along that length, and as a timing of that movement, with *real* Time still out of the picture, simply will not do. And we are being forced into a regress, where we do not want to go, chiefly by metaphors being transformed into statements of fact.

A regressive series of Times demands a similar series of Observers: Observer 1 in Time 1, Observer 2 in Time 2, and so on to infinity. (Though Dunne does in fact offer us an Ultimate Observer in Absolute Time, having had enough of infinity.) Here I must remark—as I suggested earlier—that I am suspicious of this passive role we play as Observers, for, as I said there, in our world theatre we do not feel we are members of the audience but actors; we do not spend our lives staring at the scene, *we are in it.*

Dunne knew this, of course; later he did not shirk the problem of our intervention to change the future; nevertheless, I think by turning us into Observers he made it a little easier for himself. And here I ought to make it plain that his Observers 1, 2, 3, and so on, are not different persons; they are ourselves, so to speak, in depth. Somewhere down the regress, however, higher beings must take over, preparing the way for the Ultimate Observer.

Dunne might be cautious before making a start, but once started he did not suffer from speculative timidity. What might be called his Observer Corps not only supplied look-outs for all Time dimensions; it also settled, Dunne claimed, the question of our self-consciousness:

How would you define rationally a *'self-conscious'* observer—define him so as to distinguish him from a non-self-conscious recorder such as a camera? You would begin, I imagine, by enunciating the truism that the individual in question must be aware that something which he calls 'himself' is observing. Putting this into other words, the assertion is that this 'self' and its observations are observed by the self-conscious person. But it is essential that he should observe his objective entity as something pertaining to *him*—he must be able to say: This is *my*-'self'. And that means he must be aware of a *'self'* owning the *'self' first considered.* Recognition of this second 'self' involves, for similar reasons, knowledge of a third 'self' —and so on, *ad infinitum.*

In short, we are going into the infinite regress, to keep Time company. And I for one absolutely refuse to go.

In this I am, just for once, in admired academic company. Anybody who has read Professor Gilbert Ryle's *The Concept of Mind*, a work that has been steadily reprinted since it was first published in 1949, will remember how he dismisses the idea of our being conscious of being conscious by showing us the regress trap: "And then there would be no stopping-place; there would have to be an infinite number of onion-skins of consciousness embedding any mental state or process whatsoever." I have not the space here, even supposing I had the necessary ability, to challenge Professor Ryle's various conclusions. But I will venture to say that I am extremely suspicious of them and cannot help feeling that he, like many of his philosophical colleagues now, is secretly passing ammunition to behaviorists and the more aggressive positivists. And if Dunne with his regress over-elaborates his idea of the mind, I think Professor Ryle over-simplifies it.

Because some of us can easily be caught in carefully placed linguistic traps, because we may write loosely to hurry on the argument, it does not follow that our whole conception of reality is wrong. A man who shouts "Your house is on fire" may not be able to define exactly what he means by *your* and *house* and *is* and *on* and *fire*, but he might still be saying something quite important.

Now I would agree that Dunne is working the regressive trick, like a conjuror "forcing" a card, in such sentences as "And that means he must be aware of a *'self'* owning the *'self' first considered.*" Accept this, a trapdoor flies open, and down we go into the regress. There is no such infinity of selves. Self-consciousness does not lead to an infinite regress. But even if it would seem to do, that does not mean we must rid ourselves of any notion of different selves. Whatever the theory might be, I am prepared to defend our common practice of dividing ourselves into different selves —not an infinity of them but *three* of them.

There is Self 1, which, because we are self-conscious, is always an object. There is Self 2, which may be a subject or an object. This self

observes Self 1, but can be in its turn observed by Self 3. This last self is never an object; there is no other self to observe it; we can put Self 3 on paper, as I am doing now, but we are not in practice aware of our own Self 3; it is ultimate. Let us see how the three selves work.

Self 1 says, "I disagree with Dunne and Professor Ryle." (It is saying this, so to speak, in the battle, in the welter of opinions, feelings, and prejudices.) Self 2 says, "I know that I disagree with Dunne and Professor Ryle." (It says this out of the battle, and is ready to talk or write on the subject.) Self 3 would probably say nothing, not caring a rap about Dunne and Professor Ryle; but if compelled to speak it would probably say: "I, Self 3, know that Self 2 knows that Self 1 disagrees with Dunne and Professor Ryle. And this is something I might possibly have to take into account."

In moments of great stress, Self 3 dominates the other two and takes over. Here is an actual instance that I remember. Self 1 heard terrible sounds in an aircraft, found himself hurled out of his seat and hit by a piece of luggage. Self 2 knew there was about to be a terrible accident. Self 3 said: "Well, well, I shall know in a moment what it feels like to be fried alive." Unlike the other two selves, Self 3 does not really care; it is as if it goes along with the other two just for the ride.

These three selves, however, are not in a series, except perhaps in the matter of detachment. (Self 1 cannot be detached or it could not function at all; Self 2 must have some detachment in order to acquire and sort out its knowledge; Self 3 is utterly detached, in a way that we cannot understand just because it is ultimate.) To call them a series of observers is to miss their different functions and characters. We can say, if necessary—though using the same verb over-simplifies the matter—that we know (3) that we know (2) that we are conscious (1). I do not like this way of putting it, but even so I would be prepared to defend this "knowing that we know" as an essential part of our common human experience. But any further "knowing" adds nothing. Selfhood is not regressive, but does demand three terms.

A Paris square through 24 hours: From the opening of the Métro and the first signs of life, through a busy day, to night and the closing of the Métro. Finally the water cart comes round, and another day begins. This is the chronological, one-dimensional time we are all familiar with. Though flaws can be found in Dunne's theory, he remains one of the most important of those men who insist—on the basis of their own observations—that some of our experiences belie the idea that this "kind of Time" is the only kind there is.

In his *Serial Universe*, Dunne amusingly illustrates his theory, as we see, by introducing us to a lunatic artist, who is trying to get everything he is aware of into his picture. He does not take this symbolic example very seriously but he would have been prepared to argue that once we have arrived at X_3 we must go on to X_4, X_5, X_6, and so on to infinity, whereas my contention is that reason, as distinct from lunacy, can stop at X_3. The regress adds nothing but itself. But what he writes, following the pictures, is of some importance:

The interpretation of this parable is sufficiently obvious. The artist is trying to describe in his picture a creature equipped with all the knowledge which he himself possesses, symbolising that knowledge by the picture which the pictured creature would draw. And it becomes abundantly evident that the knowledge thus pictured must always be less than the knowledge employed in making the picture. In other words, *the mind which any human science can describe can never be an adequate representation of the mind which can make that science*. And the process of correcting that inadequacy must follow the serial steps of an infinite regress.

Ignoring this last statement, which in my opinion is not necessary as a conclusion to his general argument, we ought to consider carefully his preceding statement about mind and science. It seems to me to be true and to indicate a limitation that is often conveniently forgotten. Science can function only by abstracting from the reality in which the scientist has his being. In spite of the astonishing complications it discovers, with which it dazzles and almost blinds us, science is compelled by its own terms of reference to be a drastic simplification.

Nevertheless, though reality is so rich and strange, we do not have to credit it with infinite regresses of Times and Observer-Selves. Yet if his purely metaphysical theory of Time, developed in *The Serial Universe*, was weak and very vulnerable, Dunne on his own ground was as acute in his thinking as he was original. I cannot help feeling, therefore, that his passion for the regressive, dragged in when his argument did not really need it, came from a certain mischievous element in him, from his desire as an amateur philosopher to taunt and to shock the professionals: "Ah, you wouldn't look at an infinite regress, would you? Well, watch me!" With the result that he not only invited destructive criticism but also left himself wide open to it.

But though he was so belligerently emphatic about the regressive character of his Time theory, it is significant that his more detailed accounts of how it worked were confined to the first three terms of the regress: Times 1, 2, and 3; Observers 1, 2, and 3. They were in fact all he needed.

When we met for that last time, in that ancient castle, I denounced his infinite regresses and begged him to abandon them. He did finally admit that the regresses would not do, and even suggested they were a mere appearance, a kind of mirage on the unknown desert frontier, not to be crossed by the human intellect. (Perhaps because he was host that night, he admitted so much. Commander H. E. Guerrier, who saw more of Dunne than I did, does not take him quite as far as that, but does write: "Except in theory with no practical application to any physical or spiritual experience, the unimportance of all but the beginning of the regress is insisted on in the final paragraph of Chapter 1 in his posthumous book *Intrusions*?" Just as the medieval map makers, once they had left behind them a known coastline, filled in the great blank spaces with dragons and other monsters, Dunne rushed in his regresses. They did him much harm. The professionals, who did not want their idea of Time to be challenged, spotted his weak defenses here and charged in, content afterward, when they felt they had demolished his general theory of Serialism, to ignore any genuine discovery he might have made about Time.

4.

Now we come back to what I consider to be Dunne's own true ground. Here he has discovered that in dreams—and not once in a lifetime, in the shadow of some terrible disaster, but regularly—we may see the future just as we do the past. How can this happen? How does it work?

To understand his explanation, we must grant him a Time 1 with its Observer 1, and a Time 2 with its Observer 2. What the latter observes are what Dunne calls "the brain states" of Observer 1 (his experience of Time 1)—"the sensory phenomena, memory phenomena, and trains of associative thinking" belonging to ordinary waking life. Now when Observer 1 is awake, 2 is attending to what Observer 1 is discovering in Time 1, using the three-dimensional focus with which we are all familiar.

But Observer 2 in Time 2 has a four-dimensional outlook. This means that the "future" brain states of Observer 1 may be as open to Observer 2's inspection as 1's past brain states. But it also means that when Observer 1 is asleep, his three-dimensional focus can no longer act as a guide or "traveling concentration mark" for Observer 2. The latter is now left with his four-dimensional focus, which has a wide Time 1 range—much of it "future" to Observer 1.

But Observer 2 cannot attend and concentrate properly. Now and then he may succeed, as he must do in a clear precognitive dream. Usually, however, unguided by the inactive sleeping Observer 1, Observer 2 with his four-dimensional focus is all at sea, bobbing here and there along Time 1, making nonsense of his experience.

This relation between Time 1 and Time 2, between Observer 1 and Observer 2, enables Dunne to explain why we find most of our dreams so bewildering:

Throughout your dream you endeavour to interpret the dream scenery as a succession of three-dimensional views similar to those which you experience in field 1. And always the excessive Time 1 length of your focus defeats you. Nothing stays fixed to be looked at. Everything is in a state of flux. For always your view comprises the just before and the just after of the instant of Time 1 sought for. And, because of the continual breaking down of your attempts at maintaining a concentrated focus, the dream story develops in a series of disconnected scenes. You start on a journey . . . and find yourself abruptly at the end. You are always trying to keep attention moving steadily in the direction to which you are accustomed in your waking observation—i.e. forward in Time 1—but always attention relaxes, and, when you recontract it, you find, as often as not, that it is focused on the wrong place and that you are re-observing an earlier scene in the dream story. You begin to follow up what you would recognise, were you awake, as a train of

associated images; but your attention relaxes slightly in the middle of the journey, so that what is actually perceived may be the first image in the train followed instantly by the last. That you seem to enter houses without passing through the walls is, of course, one of the most commonplace of happenings in a four-dimensional world.

Here I must interrupt Dunne to register my belief that while this is an adequate explanation of most of our dreams, especially those impossible to remember without a determined effort, it does not explain all our dreams. For there are dreams that are outside of this confusion of the three- and four-dimensional focuses—clear dreams without these jerks and disconnections, and powerfully dramatic dreams in which figures obviously symbolic play a part. I have an idea that no dreams we remember easily and never forget, no dreams in which our emotions are deeply involved, can be entirely accounted for by this relation between Observer 2 and Observer 1's "brain-states." Something else has been at work.

Dunne points out that in what he calls "the true dreams of unbroken sleep"—to distinguish them from fragmentary dreaming originating in Observer 1's bodily feelings—we never feel acute pain, "are never dazzled by bright suns, deafened by loud noises, irritated by uncomfortable garments, scorched or frozen or fatigued. Dreams," he continues, "although they seem real enough, lack all these unpleasant intensity-characteristics of waking life; we are barely aware of the presence of our bodies."

This is true but not new: The generally accepted explanation is that we avoid "these unpleasant intensity-characteristics" because in dreams the nerves that carry these signals are not functioning. But this would not serve for Dunne, who is arguing that dreams come from Observer 2's bewildered travel among the brain-states of Observer 1 (past and future in Time 1), which have had—and will have—their share of painful sensations. So Dunne's explanation, which is very ingenious, is that the intensity of such sensations depends upon the attention we concentrate upon them, and that our Observer 2, with his four-dimensional outlook, cannot pinpoint

his attention as Observer 1 can—and must—do. Our Time 2 focus is too broad and relaxed.

He says nothing here about dreams' emotional character. But we will let him continue:

Now, throughout your dreams, you *think* about that dream, just as you think about your sensory experiences in waking life. You estimate the significance of what you see in the dream; you make naive plans to cope with the dream situations; you remember what has happened immediately before in the dream. . . . It would be going too far to say that it is, in every sense, the thinking of a little child, for it involves conceptions which pertain to adult life—such as, for example, political ideas. But we may all admit that it is thinking of an extraordinarily feeble kind as compared with that which accompanies the inspection of the successive brain-states in field 1. . . .

He means there our waking experience, with our Observer 1 functioning in Time 1. And as he still says nothing about the emotional character of our dreams, I will venture to say it for him.

There are certain feelings, ranging from terror to pure joy, that come washing over us in some dreams—feelings that are unmixed as they rarely are in waking life and that remind us of our childhood. It is then that we can stumble from blind terror and various dusks and darknesses of misery to arrive without effort at great shining peaks of joy. We are never so miserably crushed or so expansively happy again—except now and then in dreams.

So it may be that in our dream-life we are not only, as Dunne suggests, thinking like children but also feeling like them. In other words, this four-dimensional Observer 2 is weak in ratiocination but has a great capacity—even though it is not always in use—for experiencing certain strong unmixed emotions. He is, we might say, considerably nearer Heaven and Hell than we seem to be in our waking life.

It is Dunne's belief—and now I am omitting some complicated technical evidence and argument, which I believe to be sound enough once his main theory is accepted—that Observer 1, with his sharp focus in Time 1, is *educating* Observer 2 and all the other attendant Observers. (And these we need not necessarily accept.) We must consider, as he says—

The interesting conception of a higher-order thinker who is *learning to interpret* what is presented to his notice, the educative process involved being his following, during the waking hours, with unremitting, three-dimensional attention, the facile automatic action of that marvellous piece of associative machinery, the brain. . . . Whatever capacities for eventually superior intelligence may be latent in the higher-order observer, they are capacities which await development. At the outset brain is the teacher and mind the pupil. Mind begins its struggle towards structure and individuality by moulding itself upon brain. Evolution has worked for possibly eight hundred million years towards the development of brain. . . . And it now appears that, apart from its self-sustaining and self-developing activities, the brain serves as a machine for teaching the embryonic soul to think. . . .

This is, we must admit, a bold and brilliant conception. It begins to reconcile two formidable opponents, belonging to two opposing schools of thought, the religious and the scientific-evolutionary. It offers, at a first glance, a wonderful way out. Once again, however, I feel that Dunne's position is undermined and so weakened by his obstinate regressing thinking. (I apologize here for the absence of quotations in what follows but I simply have not room for them. I can only hope I am not doing an injustice to a man who cannot reply.)

All the confusion and trouble come from his having saddled himself with an infinite regress of Observers, presumably belonging to higher and higher "orders." For sometimes he writes as if, somewhere at the unimaginable end of the line, there is an all-wise and omnipotent Ultimate Observer, another name for God. At other times, as in the last quotation from him, he refers to a self-educating process, with successive Times like higher forms in a school, for the individual mind or soul. It is possible to believe—we are now on familiar ground—that God is educating us; it is also possible, though harder, to believe that we are educating God; but I find it impossible to understand what is happening further along this endless vista of Times and Observers. But then I do not believe it exists. The regress is all wrong.

Now we come to the hard part. Let me put it briefly and brutally. The future can be seen, and

because it can be seen, it can be changed. But if it can be seen and yet be changed, it is neither solidly there, laid out for us to experience moment after moment, nor is it non-existent, something we are helping to create, moment after moment. If it does not exist, it cannot be seen; if it is solidly set and fixed, then it cannot be changed. What is this future that is sufficiently established to be observed and perhaps experienced, and yet can allow itself to be altered?

I propose to take an actual example of "interference" and explain it as best I can in Dunne's terms. Let us take the case of the American mother who dreamed she visited a certain creek, left her baby because she had to fetch some soap, then returned to find her baby dead (p. 225). When she found herself in waking life in this exact situation, she remembered her dream and suddenly changed her mind about leaving the baby. I will call her Mrs. X. And as Observer 1 in Time 1, she will simply be Mrs. X1, and as Observer 2 in Time 2 she will be Mrs. X2.

Now while Mrs. X1 is asleep, Mrs. X2 happens to see what lies ahead in Time 1—visit to the creek, fetching the soap, death of baby. This experience becomes, for Mrs. X1 on waking, a remembered dream. But Mrs. X1 has many other things to attend to as her pin-pointed *Now* travels along the fourth dimension. Finally, no longer remembering the dream, Mrs. X1 (who is also of course Mrs. X2 as well, mentally but not physically) arrives at the creek, puts down the baby, and prepares to wash the clothes. Just as she is about to leave the baby, Mrs. X2, we might say, intervenes, and Mrs. X1 picks up the baby; and at that point what Mrs. X2 saw as the Time 1 future for Mrs. X1 *is changed*. By intervention from Time 2 and action in Time 1, at the *Now* point where such action is possible, what will happen in Time 1 is quite different, a baby who was going to die being rescued and eventually growing up to manhood.

(Though it takes us away from our immediate subject and returns us to one dealt with earlier, I must make the point here that Mrs. X's dream cannot reasonably be accepted simply as a warning, with the Time element left out. She was not telling herself in the dream that *if* she

went to one of those creeks her friends had mentioned, *if* she decided to do some washing and to go for the soap, and *if* she foolishly thought it safe to leave the baby alone, then the baby might wander into the water and drown. These *ifs* and a vague anxiety could not possibly create the numerous vivid details she recognized later at the creek. So either Mrs. X was deceiving herself and adding a few lies or she was genuinely involved in a precognitive Time experience. And for my part I do not see why we should accuse this lady and hundreds of others in order that the conventional idea of Time should remain undisturbed.)

This intervention from Time 2 leading to an action in Time 1, changing the Time 1 future, seems to me to offer us a chance of ending the old quarrel between free will and determinism. Its opposed views cannot be reconciled if we assume our existence to be on one level. Along the Time 1 length, without any intervention from Time 2, our life may be said to be determined. But if our Observer 1 in that time does not ignore the intervention of Observer 2 from Time 2, then we may also be said to enjoy some measure of free will, accepting an offered choice.

We can also explain—in these Dunne terms—the bewildering future-influencing-present (FIP) effect, which made its appearance in section 2 of the chapter "The Letters." So let us return to Dr. A and Mrs. B (p. 201). Now Dr. A fell in love with Mrs. B and married her. These events must have been responsible for some excited "brain-states" along his Time 1 length. His Observer 2, wandering in his four-dimensional fashion, must have attended to and picked up this excitement. This might have resulted in a precognitive dream, but the absence of it does not mean a total lack of communication from Time 2 to Time 1. As we know, the mere sight of the unknown Mrs. B's name, even on official notices, not personal letters, made Dr. A feel an extraordinary excitement: It was a present from himself in Time 2 to himself in Time 1.

So far we have not reached the really hard part. The great puzzle now looms up. In Chapter XXV of his *Experiment with Time*, Dunne begins by asking the right question:

Since observer 2 sees what lies 'ahead' (in Time 1) of observer 1, there must be something there for him to see. And whatever is thus positively there must be 'pre-determined' from the point of view of observer 1 employing Time 1. So the question arises: What is this 'something', and how is observer 1 able to alter it by intervention at his Time 1 'now'?

Then throughout the rest of this short chapter, he makes some impressive-looking statements about any world described by observation being relative to the describing observer, about our being always outside any world of which we can make a coherent mental picture, about the physical world that we describe exhibiting itself as deterministic in Time 1 but having "a contact point" with our Observer 1 at the traveling Time 1 "now." But he does not supply us with an answer to that all-important question—*What is this "something"?*

Let us bring back Mrs. X and her baby. And once again we call her Mrs. X1 in Time 1, Mrs. X2 in Time 2. Her dream of the dead baby, Dunne would tell us, comes from Mrs. X2's attention to one of Mrs. X1's "future" brain-states. In Time 1 Mrs. X1 will see her baby dead; in Time 2 she is already seeing it, hence the dream; then there is the intervention that has already been described.

But what is this dead baby? On Dunne's theory it has to be more than an image because Mrs. X1, as distinct from Mrs. X2 or Mrs. X3, cannot *imagine* a baby. Mrs. X1, according to him, provides the sharp three-dimensional focus traveling with its *Now* along the Time 1 length, together with any ratiocination that may be necessary: She is in the recording business, really working for Mrs. X2, not in the imagination department.

Therefore if there was a dead baby in those Time 1 future brain-states of hers, then there was a *real* dead baby, perhaps 20 pounds of him, to be observed by that three-dimensional focus. And if there was, then what became of him when Mrs. X1, prompted by Mrs. X2, took an action that changed her future? In what world did the baby die? I for one cannot escape this question, and to my mind Dunne supplies no answer to it. And unfortunately it remains the *key* question.

It is true that Dunne suggests later, as far as I can follow him, that it is mere "materialism" to insist upon a physical world existing independently of our pictures of it, our brain states. But once he adopts this extreme attitude of metaphysical idealism, he seems to bring crashing down his whole theory of the relations between his regressive observers, who are all being educated by the Time-limited but sharply-focused Observer 1, with his slow but sure three-dimensional experience.

What he is experiencing in Time 1 is our existence in the physical world, also in Time 1. We need it, he suggests, for our education, like the toys in a kindergarten. We have to make our start on this clumsy and limited physical level. But then when he no longer needs it, I feel that Dunne not only makes it vanish but also pretends it never was there. And he leaves me still asking: *What about the dead baby?*

5.

Readers who do not know Dunne and wish to avoid the formidable intricacies of *An Experiment with Time* and *The Serial Universe* should begin with two later and shorter books, intended for a popular audience. (Ironically enough, however, it is *An Experiment with Time* that has to be regularly reprinted.) These two later and shorter books are *The New Immortality*, first published in 1938, and *Nothing Dies*, published in 1940. Although both of them include a simplified account of his general theory of Serialism, as their titles suggest, their main emphasis is upon Dunne's ideas of life after death.

We are, he insists, immortal beings. It is true that we "die" in Time 1 when our Observer 1 reaches the end of his journey along the fourth dimension. And then all possibility of intervention and action in Time 1 comes to an end. This limits Observer 2's experience (through Observer 1's brain-states) of Time 1, but it does not involve the death of Observer 2, who exists in Time 2. No longer having any Time 1 experience to attend to—we must remember that this is what Observer 2 does, except when Observer 1 is asleep or relaxing his attention—Observer 2 now experiences Time 2 as Observer 1 did Time

1: That is, it is for him ordinary successive time as we know it. He has to begin learning all over again as his four-dimensional focus moves along the fifth dimension or Time 3. People and things will be the same and yet not the same. We catch glimpses, though confused and distorted, of this after-death mode of existence in our dreams.

The four-dimensional focus of Observer 2 in Time 2, as we have seen, produces the confusion and distortion of dreams. He lacks the clear three-dimensional focus and rational guidance of Observer 1, who is asleep. But this Observer 2, as Dunne points out, is able to "blend two or more sensory impressions widely separated in Time 1 into a single but more complex impression." (Example: Observer 1 sees a blue dress in a shop window, then later a certain girl elsewhere; Observer 2 can combine these images into one, the girl now wearing the blue dress. And certainly our dreams often show us this blending of images.) He suggests that in our Time 2 "after-life," once we have understood how to live it, we shall be able to blend, combine, build, with all the elements of our Time 1 existence, using them more or less as a composer does his notes, an artist his paints.

He also suggests what will happen to our personal relationships (in the following quotation, please note, the references in brackets are mine):

In the world of what we may call the lesser 'now' [Time 1] a meeting is a state where the bodies of two people are in close proximity. The attentions of both persons are focussed on that instant of pseudo-time [Time 1], and the communication between mind and mind which is the essence of the meeting follows through the ordinary media of speech or signal. In the greater 'now' [Time 2 but now, after death, Observer 2's Time 1], your attention may visit a scene, and you may see again the one you seek. You may hear again the spoken words, you may receive and give the same caresses. But the attention of that other may not be there. In that case there is no *meeting*. Moreover, in the world of the greater 'now', communication is not by word or gesture, but (as I shall show later) through the medium of a common field of consciousness. Mind communicates direct with mind . . . meeting in that world requires, if it is to last for more than a fleeting instant, mutual desire. But it requires something more than that if you are to

savour it in full perfection. That is where ethics come in. Bear in mind, please, to begin with, that the one you seek is engaged in his or her world-building, and that the edifice aimed at is fairly certain to differ in many essentials from that which you would plan. Now, you can be a little *god* in your own little kingdom. You can make everything happen exactly as you please. You can meet again every one you have ever known, at any age you can remember. They will welcome you gladly—if you wish it. They will acknowledge that you had been right, after all, in those little quarrels. But, presently, unless you are beyond measure foolish, you will realise that this docility does not ring true. It will be a terrible moment when you discover that the words are dictated by you: that the affection is of your own inventing. You will have what you have wanted always in this life—a world wherein every wish is fulfilled. It is a little heaven of private pleasure—and a hell of utter loneliness. To avoid, or to escape from, that, you must be willing to surrender some of your sovereignty. You must be prepared to build to please others. . . .

Simple well-tried ethics—and no worse for that —but set against a background that seems to me an impressive intuitive or imaginative effort on the part of an elderly aeronautical engineer.

His posthumous book, *Intrusions?* (which was written, as Mrs. Dunne tells us, "with a sense of urgency and in circumstances of chronic ill-health," never revised, never even completed), is more personal in its matter and manner, more religious in tone, than the earlier and more important books. There is one passage toward the end, not directly connected with the Time problem but not entirely separated from it, that will serve as our final quotation from Dunne:

The flow of nervous energy in any system of nervous matter tends to follow paths of least resistance. A flow of that kind is called, of course, 'automatic'. Such automatic flows tend to become (in the life-history of the *genus* concerned) uncontrollable and unconscious. . . . *Allow a specific automatic activity to function for many generations without frequent interference and it will become first uncontrollable and then unconscious.*

Now then: picture a civilization in which all stress, danger and difficulty in meeting daily needs have long disappeared; in which adherence to a code of mutual social respect coupled with mutual social compliance has become second nature to all, so that no individual suffers from lack of companionship or from rejected emotional cravings; a world in which music, art and entertainment are all State-provided and in accordance with the demands of the mediocre majority; a world in which 'flaming souls' would be shocking anachronisms. What would happen?

Man's marvellous brain would, of course, adapt itself swiftly to the new stereotyped conditions. Its response to each recurring situation would be perfect, and leave nothing for the mind to do. Mind could rest. It could sit placidly admiring the sycophantic activities of its mechanical servitor, the brain. There would be no longer the smallest need for mind's interference. Let the Human Race continue in this condition without interruption from a conquered Nature. What, in the long run, would be the result?

Man would become an almost entirely unconscious automaton.

To this we might add that not only would men become almost like machines, but also that they would soon regard themselves (there are signs of this now among men who proudly introduce us to their computers and other electronic devices) as *inferior* machines. The superior machines would take over. But never if men understood that their minds existed in times no machine could know.

Now I repeat in effect what I said at the beginning of this chapter—that those of us in any way concerned with the Time problem owe a great debt to John William Dunne. We may reject, as I do, his regressive notions of Time and personality; we may believe that the professional philosophers defeated him when he fought on their ground for his metaphysical proofs of the regress of Time; we may feel, again as I do, that he wasted too many years and too much energy chasing hares that he himself had brought into the scene. But his experiments with dreams and his analysis of the dreaming self seem to me to be worth more than anything produced by the combined efforts of all other people in this field.

Nobody now can seriously examine the Time problem without taking Dunne's work into account. It was never as widely recognized and appreciated as it should have been, during his lifetime. He has still to be praised and honored as one of our great originals and liberators. I should like to think there was some way of conveying these thanks and a loud and affectionate *Bravo* to his Observer Two or Three.

11   Esoteric School

11 Esoteric School

I.

The elaborate system of thought, behavior, psychological development, taught by Gurdjieff and his chief disciple, Ouspensky, was often called by them and their pupils *the Work*. To save space and trouble I shall follow their example. Now, to begin with, it is surprising how little public attention has been given to the Work. A good deal has been written about it from the inside—I must possess at least 20 of such books myself—but, so far as I know, nothing of importance from the outside. If a disinterested critical examination of Gurdjieff's teaching and ideas exists it has never come my way. There are two reasons why I find this neglect surprising.

The first is that from the early 1920s onward, groups dedicated to studying the Work came into existence in Paris and London and probably other European capitals, in New York and Mexico City and various places in South America. (As the Work had no central organization and never advertised itself—a fact worth remembering—I doubt if a complete list of its groups is available anywhere.) The second reason, though here I can refer only to the British groups, is that although their numbers may not have been impressive, their quality was. Any idea that this was a movement supported by rich foolish women, for the benefit of charlatans, can be dismissed at once.

No doubt some rather dim people did drift in and out of the Work. But to name only two of its students, now dead: A. R. Orage, once the most brilliant editor in Britain, and Maurice Nicoll, a pupil of Jung and then a distinguished Harley Street specialist, were anything but dim. And indeed the Work was always able to provide itself with suitable buildings because it could command the services of architects, engineers, expert professional men. The level of Gurdjieff's and Ouspensky's most devoted students was very high. In order to study this movement, nobody will have to do any intellectual slumming.

I make these points simply out of a sense of fairness. I was never a member of any one of these groups myself; I never set eyes on Gurdjieff or Ouspensky; and my acquaintance with their chief successors in the Work has been very limited. But a certain amount of rubbish, deposited by ignorance and prejudice, must be cleared away. True, the Work has a semi-oriental background, arriving from mysterious sources in mid-Asia, where Gurdjieff (himself a kind of humorous magician who enjoyed elaborate mystifications) was supposed to have spent years in search of esoteric truths, ancient secret wisdom. But the Work is far removed from the usual soft and sentimental doctrines of Higher Thought, Theosophy, and the rest: It is hard, demanding, grimly unsentimental.

It insists upon men making unwearied efforts to free themselves from a waking sleep or being mere machines, to become fully conscious, to build up a central commanding "I" in place of a score of contradictory "I's," to rid themselves of wasteful and stupid negative emotions, to make "essence" grow at the expense of false "personality," and not to imagine they are in easy possession of immortal souls but to believe that in the end, after unremittent effort, they might create in themselves such an indestructible soul. (It is in fact a kind of esoteric Christianity, and readers interested in this aspect should look at Maurice Nicoll's *The New Man* and *The Mark*, in which he re-interprets the Gospels.)

I wish there were space here for some account of Gurdjieff's psychological system and his elaborate and fascinating cosmology, but our subject is Time and we must pursue it. I will only add—as an outsider who is merely a reader and was never a member of the personally instructed groups (and this was considered to be essential) —that while some of the Work seems to me as dubious as Gurdjieff's claim to have rescued this "ancient wisdom" from unnamed remote mid-Asian monasteries, he does often reveal some

astonishingly original insights, and some of the simpler "work on oneself" does actually bring some excellent results. Whether Gurdjieff was a new prophet and teacher or a Near Eastern original, two thirds genius, one third charlatan, he certainly knew a great deal more about our common humanity than most of us know.

However, we must return to Time. This brings us to Ouspensky. He was a Russian journalist, author, lecturer, who had some acquaintance with the sciences and mathematics but was chiefly interested in what he called "the miraculous" and in the possibility of discovering some esoteric "school" (a very important term afterward in the Work) in which initiates would receive personal teaching. (Years later, when he lectured to groups in London, the proceedings were given an air, very Russian, of quite unnecessary mystery.) He had just finished a long journey in the East when the First War broke out:

I said to myself then that the war must be looked upon as one of those generally catastrophic conditions of life in the midst of which we have to live and work, and seek answers to our questions and doubts. The *war*, the great European war, in the possibility of which I had not wanted to believe and the reality of which I did not for a long time wish to acknowledge, had become a fact. We *were in it* and I saw that it must be taken as a great *memento mori* showing that hurry was necessary and that it was impossible to believe in "life" which led nowhere. . . .

In Moscow in 1915 he met Gurdjieff, who had gone to Russia to teach some groups, and he was soon under the spell of that powerful and enigmatic personality. They were obviously very different types. Ouspensky bookish and solemn, Gurdjieff vastly more experienced, with a richer and warmer nature, sly and humorous, an Autolycus turned sorcerer and sage. When, after the Russian Revolution and during the Civil War, Gurdjieff had to lead his party of followers through the wilds of the Caucasus, where Reds and White were fighting, he coolly announced

that they were about to do some gold mining in the neighborhood, wherever they happened to be, and so they were unmolested, often helped, by both armies. (With people in authority or inquirers he did not welcome, Gurdjieff liked to appear rather simple and stupid, some old carpet merchant too far from home. There is an innocently revealing account of this performance by Rom Landau in his *God is My Adventure*.)

Ouspensky soon accepted the authority of Gurdjieff's teaching without question, and became its leading exponent. But gradually he made a distinction between Gurdjieff the man and the Work itself. By the time, in 1922, Gurdjieff had opened his famous Institute at Fontainebleau, Ouspensky had already settled in London, at work with his own groups. He remained in London for nearly 20 years, going to America during the war and then returning afterward to England, where he died in 1947.

(It has been noticed that the leading personages in these Gurdjieff-Ouspensky groups, so determinedly devoted to self-development, were by no means free of misunderstandings, acrimonious disputes, downright quarrels. These could not have been much helped by the chilly reserve and severe authoritarianism of Ouspensky, who probably never felt at home in England and denied himself to all but a few of his closest associates. It is only fair to add, however, that the Work itself, after a long period of unquestioning pupilhood, encouraged men to strike out for independence; and Gurdjieff's insults and sudden rages were a part of his tactics, compelling his chief followers to stand up for themselves and march away, to make what they could out of his system, free of his direct influence. Here was a very remarkable man.)

Something has had to be said about Gurdjieff because of his influence upon Ouspensky. It is Ouspensky, however, who is important here, just because it is he and not Gurdjieff who is concerned with the Time problem. Gurdjieff—and

I say this only on the basis of what I have read—never appears to have had this concern. He seems to have been rather evasive, neither fully agreeing with Ouspensky's theories nor completely denying them. But while Time theories as such played no part in the Work, the "Merciless Heropass," which has a sinister cosmological role in his *All and Everything*, may be regarded as symbolic of Time. Moreover, we are told that the Work aims at freeing us from the destructive action of Time "by reproducing in ourselves the creative act whereby the Universe itself is liberated from destruction" (Bennett).

Ouspensky's interest in the subject was there long before he met Gurdjieff. It can be discovered in *Tertium Organum*, published in 1922, and his first book to be translated into English. Though it has been frequently reprinted, it is to my mind much inferior to his second book, published in 1931, *A New Model of the Universe*. In his "Prefatory Note" Ouspensky claims that this book was begun and practically completed before 1914. But the chapters are dated, and what is for us the all-important chapter, "Eternal Recurrence," is dated 1912–1929, and by 1929 Ouspensky had not only accepted Gurdjieff's system, being indeed now its chief exponent, but had also more or less broken with his old master. So whether this is pure Ouspensky, brought up-to-date from 1912, or Ouspensky fortified by the brandy of Gurdjieff, I do not know. What I do know is that no examination of the Time problem can afford to let this chapter on "Eternal Recurrence" go unnoticed. Its bold challenge must be faced: Either it must be accepted or rejected.

2.

Ouspensky's theory of eternal recurrence must not be confused with the Great Circle or Year of the ancient Greeks, which I described in Chapter 5 (in Part Two). And it has nothing to do with Nietzsche's idea of recurrence, which Ouspensky, in a footnote, sharply and rightly rejects. (Nietzsche argued that the illimitable resources of the universe must produce other earths exactly like this one, and that then the same causes will make everything recur. But this is not true be-cause the odds against exact repetition will multiply faster than the chances of it.) Ouspensky quotes examples of Western thought and feeling that suggest the idea of recurrence, but they are not impressive. After all, the *déjà vu* effect is fairly common, and, as we have seen, there are explanations of it that do not involve recurrence.

Ouspensky's theory of eternal recurrence is really an extension—and, in my opinion, by no means a necessary one—of his idea of three-dimensional time, explained in a previous chapter, called, like the book itself, "A New Model of the Universe." Because, to present this idea, I shall have to quote his text, as I shall also have to do when we come to his theory of recurrence, there is a point I should like to make first.

It is this. Though I consider Ouspensky to have been a man of considerable intellect who produced genuinely original ideas of some value, and not at all the charlatan many scientists might imagine him to be, I think we are entitled to entertain some suspicion of his style and manner. A good scientist can be allowed a style and manner of dogmatic certainty, because he will offer proofs of what he is stating. Where he is uncertain, because no proofs are to hand, he will make plain his uncertainty.

On the other hand, I believe there is no harm in a man being freely and frankly speculative, saying in effect "I cannot help feeling" or "I have come to believe" whatever it may be. He takes the risk, it is true, of being denounced by logical positivists for writing emotive bosh. But that is no reason why he should shrink from exploring the odd corners of his mind or trying to understand his queerer experiences. If in doing so he is merely being speculative, not making statements worth serious consideration according to logical positivism, then very well, there it is. (Without some license for speculation, advanced science might soon grind to a halt. And as Professor H. H. Price says: "If such writers as Hume and Mach and the modern Logical Positivists had lived in the early seventeenth century, physics would never have got itself started.")

But I agree with the other side in thinking than a man must be either conscientiously scientific in his approach or frankly speculative.

What he must not do is to present his speculative untested ideas in a manner and style of dogmatic certainty that they are not entitled to claim. This is being pseudo-scientific. And with all his exceptional qualities, Ouspensky, like one or two of his disciples, too often assumes this pseudo-scientific manner, as if all were proved when in fact some things have not been proved, only asserted. This does not mean that his ideas are not worth serious consideration; I believe they are; but we must be a little suspicious of a manner that demands unquestioning acceptance.

Before arriving at eternal recurrence, Ouspensky declares that Time is three-dimensional:

The three-dimensionality of time is completely analogous to the three-dimensionality of space. We do not measure space by *cubes*, we measure it linearly in different directions, and we do exactly the same with time, although in time we can measure only two coordinates out of three, namely the duration and the velocity; the direction of time is not for us a quantity but an absolute condition. Another difference is that in the case of space we realise that we are dealing with a three-dimensional continuum, whereas in the case of time we do not realise it. But, as has been said already, if we attempt to unite the three coordinates of time into one whole, we shall obtain a spiral.

After referring to the "excessive complexity" of "relativism," which presumably means the theory of relativity, a complexity resulting from its attempt to squeeze the universe into four dimensions, he continues:

The three dimensions of time can be regarded as the continuation of the dimensions of space, i.e. as the "fourth", the "fifth" and "sixth" dimensions of space. A "six-dimensional" space is undoubtedly a "Euclidean continuum", but of properties and forms totally incomprehensible to us. The six-dimensional form of a body is inconceivable for us, and if we were able to apprehend it with our senses we should undoubtedly see and feel it as three-dimensional. Three-dimensionality is the function of our senses. Time is the boundary of our senses. Six-dimensional space is reality, the world as it is. This reality we perceive only through the slit of our senses, touch and vision, and define as three-dimensional space, ascribing to it Euclidean properties. Every six-dimensional body becomes for us a three-dimensional body *existing* in time, and the properties of the fifth and the sixth dimensions remain for us imperceptible.

Six dimensions constitute a "period", beyond which there can be nothing except the repetition of the same period on a different scale. The period of dimensions is limited at one end by the point, and at the other by infinity of space multiplied by infinity of time, which in ancient symbolism was represented by two inter-section triangles, or a six-pointed star.

He goes on to examine these three dimensions of Time considered as the fourth, fifth, and sixth dimensions of space. The before-and-after line of the fourth dimension, the time we all recognize, needs no explanation. The fifth dimension forms a surface in relation to this line. Along it is the perpetual *now* of any given moment. In this fifth dimension is the true eternity, not unending movement along the fourth dimension but all the perpetual *nows*. But what about the sixth dimension, the third dimension of Time?

The sixth dimension will be the line of the actualisation of other possibilities which were contained in the preceding moment but were not actualised in "time". In every moment and at every point of the three-dimensional world there are a certain number of possibilities; in "time", that is, in the fourth dimension, one possibility is actualised every moment, and these actualised possibilities are laid out, one beside the other, in the fifth dimension. The line of time, repeated infinitely in eternity, leaves at every point unactualised possibilities. But these possibilities which have not been actualised in one time, are actualised in the sixth dimension, which is an aggregate of "all times". The lines of the fifth dimension, which go perpendicular to the line of "time" in all possible directions, form the solid or three-dimensional continuum of time, of which we know only one dimension. We are one-dimensional beings in relation to time. Because of this we do not see parallel time or parallel times; . . . we do not see the angles and turns of time, but see time as a straight line.

Until now we have taken all the lines of the fourth, the fifth and the sixth dimensions as straight lines, as coordinates. But we must remember that these straight lines cannot be regarded as really existing. They are merely an imaginary system of coordinates for determining the spiral.

Now I shall presently begin disagreeing with Ouspensky, but we have not arrived yet at that point. No doubt many readers are already dismissing this account of a sixth dimension, with its actualizing of all possibilities, as complicated semi-mystical nonsense. But before they flip over

the next pages in disgust, I ask them to remember that dream of the dead baby in the creek. The death of that baby could be an unactualized possibility, never happening in the fourth dimension, in chronological time. But the dream cannot be dismissed as a mere nothing; it may have saved a baby's life. And not being a mere nothing, it existed somewhere, in its own place and time.

In other words, I believe that a full explanation of that dream and its consequences may possibly demand a three-dimensional Time. Ouspensky here may have been right, he may have been wrong, but nobody should be in a hurry to accuse him of talking nonsense.

I think he is wrong, however, when he conjures out of those three dimensions his theory of eternal recurrence. Ignoring the spiral, now using a wave analogy, he argues that the line of life or "time" moves in a curve and makes a complete revolution, coming back to the point of its departure. The moment of death coincides with the moment of birth. A man dies only to be born again. From this, to which I cannot give any assent, he goes on to make several points I have long felt to be true:

Life in itself is *time* for man. For man there is not and cannot be any other time outside the time of his life. *Man is his life.* His life is his time.

The way of measuring time, *for all*, by means of such phenomena as the apparent or real movement of the sun or the moon, is comprehensible as being convenient, practically. But it is generally forgotten that this is only a formal time accepted by common agreement. Absolute time for man is his life. There can be no other time outside this time.

If I die today, tomorrow will not exist *for me*. But, as has been said before, all theories of the future life, of existence after death, of reincarnation, etc., contain one obvious mistake. They are all based on the usual understanding of time, that is, on the idea that *tomorrow* will exist after death. In reality it is just in this that life differs from death. Man dies because his time ends. There can be no tomorrow after death. But all usual conceptions of the "future life" require the existence of "tomorrow". What future life can there be, if it suddenly appears there is no future, no "tomorrow", no time, no "after"? Spiritualists . . . and others who know everything about the future life may find themselves in a very strange situation if the fact is realised that no "after" exists.

The reason why I agree with this limitation will appear later; but this agreement does not involve any acceptance of the idea of eternal recurrence. Having argued that eternity demands repetition, one life and time ending and another beginning, Ouspensky explains exactly what he has in mind:

This means that if a man was born in 1877 and died in 1912, then, having died, he finds himself again in 1877 and must live the same life all over again. In dying, in completing the circle of life, he enters the same life from the other end. He is born again in the same town, in the same street, of the same parents, in the same year and on the same day. He will have the same brothers and sisters, the same uncles and aunts, the same toys, the same kittens, the same friends, the same women. He will make the same mistakes, laugh and cry in the same way, rejoice and suffer in the same way. And when the time comes he will die in exactly the same way as he did before, and again at the moment of his death it will be as though all the clocks were put back to 7.35 a.m. on the 2nd September 1877, and from this moment started again with their usual movement.

But after making this declaration, Ouspensky modifies it. Only people of "deeply-rooted, petrified, routine life" will live exactly the same life over and over again. There are three other types.

There are the outwardly successful who will find it easier and easier to succeed. There are "people whose life contains an inner ascending line, which gradually leads them out of the circle of eternal repetition and causes them to pass to another plane of being." And then there are the types with a "growing tendency to degeneration," who with each new life sink lower and finally cease to be born, for "souls are born and die just like bodies." A soul may die on one plane of being and pass to a higher plane, or it may die altogether, vanishing and ceasing to be.

Ouspensky embroiders this theme with great skill; and it all makes fascinating reading. I have no space here even to begin doing justice to his astonishing play of ideas, notably his strange but not absurd notion that reincarnation does not go forward into the future but back into the past—a great freed soul takes the place of somebody in the past in order to help humanity, whose past (which is still existing and so is still shaping

our present and future) must be changed if men are to know a better present and future.

But though I have referred to Ouspensky's play of ideas, he was not in fact playing with them; however fantastic they may seem, he was entirely convinced of their truth. We are told how, during the last few weeks of his life, when he was often in great pain, by a formidable effort of will he insisted upon revisiting the places where he had lived in and around London. Presumably he was trying to fix them so deeply in his memory that when his life recurred *he would remember them.*

(We are told this by one of his closest and most devoted admirers in these last years, Rodney Collin Smith, who after Ouspensky's death organized Work groups in Mexico City and elsewhere in Mexico. In 1956 he visited Peru. and after climbing to the cathedral belfry at Cuzco he had a heart attack, fell to the street, and was killed. He was a writer as well as a teacher of the Work, and under the name of Rodney Collin published *The Theory of Celestial Influence* and *The Theory of Eternal Life.* These are attractively written and offer many clues to Ouspensky's later thinking—Ouspensky being the master and the author the most devoted of his pupils—but they cover too much ground too quickly and are too obvious examples of the pseudo-scientific approach, as if the general reader were an admiring member of a Work group.)

I believe every intelligent reader should make the acquaintance of Ouspensky's *A New Model of the Universe.* And perhaps the most fascinating chapter in it is that on "Eternal Recurrence." Nevertheless, though Ouspensky employs here all the formidable resources he can command, he cannot bring the idea home. In the way he presents it, his eternal recurrence simply will not work. If he had suggested that after death, outside chronological time, men might believe (as drowning men are said to believe) that they were living their lives all over again, then that, to use two of his own favorite terms, would have been a possibility, whereas his eternal recurrence is an impossibility. As soon as changes are introduced, with some men rising and others falling, the exact recurrence of a time is not taking place.

The complete repetition, with which he begins, is destroyed by the elaboration of his theory.

What he offers us finally is not recurrence and it is not eternal. Let us say with him that it is easier and easier for Mozart to compose masterpieces and for Napoleon to win battles, because they have been doing it over and over again. But how did they do it the first time? And when was the first time? I am writing these words on a fine Friday morning. How often have I sat at this desk on this particular Friday morning? Is this the 10th, the 100th, the 1000th occasion? In a much later book, *The Fourth Way,* not written by Ouspensky but made up of transcriptions of questions put to him and his answers, he is far less certain and dogmatic that he is in *A New Model of the Universe*:

When we think about recurrence, we think that everything repeats, and this is exactly what spoils our approach to it. The first thing to understand about recurrence is that it is not eternal. It sounds absurd, but actually it is so, because it is so different in different cases. Even if we take it theoretically, if we take purely people in mechanical life, even their lives change. Only certain people, in quite frozen conditions of life, have their lives repeating in exactly the same way, maybe for a long time. In other cases, even in ordinary mechanical life, things change. If people are not so definitely governed by circumstances, like great men who have to be great men again and nobody can do anything about it, there are variations, but again not for ever. Never think that anything is for ever. It is a very strange thing, but it seems as though people who have no possibilities, either owing to certain conditions, or to their own insufficient development, or to some pathological state, can have their lives repeating without any particular change, whereas in the case of people with theoretical possibilities, their lives can reach certain points at which they either meet with some possibility of development or they begin to go down. . . .

Nothing is said here, as it was in the earlier book, about ordinary chronological time endlessly repeating itself. With all these changes taking place, this would be impossible. We seem to be concerned with an illimitable number of private times, repetitions and variations of life, removed from history and the time-track of this earth.

I strike a match and when my pipe is alight I put the charred matchstick into the ashtray. Now

I can imagine various mental states in which I keep on lighting my pipe and tossing used matches into ashtrays. But—and now I join the scientists on the other side—I cannot believe in a chronological time, no matter how it moves in waves and circles, that restores to my hand the match that has been entirely changed by combustion. That match, like the hand that made use of it or the tobacco it ignited, belongs to irreversible time, entropy time, world time, actualization-of-one-possibility time. Not all our theorizing, not the grandest conference of esoteric schools and magicians fetched from the remotest of mid-Asian monasteries, can restore the molecules of that match to their original order. Its appearance, yes; its actuality as a physical object, related to other objects, no.

What Ouspensky describes as happening might possibly happen in what we call Time 2. But in Time 1, with which he begins and then conveniently forgets, it could not happen. Ouspensky's originality and his breadth of intellect cannot be denied. These are best displayed, however, not in his theory of an eternal recurrence that is neither eternal nor truly recurrent, but in the chapter rightly entitled "A New Model of the Universe," in which he discovers that Time may be said to be three-dimensional.

One last point. It is true, as Ouspensky points out, that many people cannot help feeling, in certain situations and at odd moments, that all that is happening has happened before. But I for one am not impressed by this as an argument in favor of recurrence. The neurologists can explain some of it, and Dunne, with his theory of precognition and prevision in dreams that we often forget, can explain the rest. And when the argument is from my experience, then a man must speak for himself. My own experience supports Dunne and the neurologists but not Ouspensky. I have never caught myself—and I am fairly adept at catching myself—feeling such a strange sudden familiarity with all that was happening around me that I also felt that this was a life I had already lived, that Time was repeating itself. That may mean, in Ouspensky terms, that I am one of the stupid forgetful types doomed to go round and round writing over and over again that I am one of the stupid forgetful types. It may also mean he was wrong.

3.

I have already mentioned, as one of the teachers of the Work, Maurice Nicoll. Readers who wish to learn how he taught it should read his several large volumes entitled *Psychological Commentaries on the Teaching of Gurdjieff and Ouspensky*, which contain the actual lectures he gave to members of his groups. They can be read with interest even by those who reject their main ideas. Nicoll's early training was entirely scientific and psychological; he took a first in science at Cambridge, went to St. Bartholomew's Hospital (the famous "Bart's"), then pursued his studies in Vienna, Berlin, and Zurich. This was before the First War, in which he served as a medical officer. He met Ouspensky in 1921, and the next year joined Gurdjieff at the Institute at Fontainebleau. Returning to London, he combined a successful practice as a specialist with studying and, from 1931, teaching the Work.

I offer these biographical details to show that a scientific training and background do not forbid a man to reject the conventional idea of Time. Indeed, Nicoll's *Living Time* is a plea, eloquent and enriched by quotations from a wide range of authors, for our self-deliverance from the domination of "passing time"—a term that I borrowed from Nicoll earlier in this book. There is no attempt to discuss the structure and possible mechanics of Time. Nicoll's approach is psychological and philosophical, as we should expect from his own background and training.

Like the mystics whom he frequently quotes, he insists upon the supremacy of the invisible over the visible, the inner world over the outer, what is discovered by the searching mind over what is revealed to the senses. The tyranny of the idea of passing time allows no broadening and heightening of consciousness. We inhabit a small portion of what he calls "living Time"; it is the belief that only our senses can show us truth that turns all this richness of life into "passing-time and death and destruction."

He goes on: "From other parts of living Time we receive a few signals—in literature, architec-

ture, art." All history, he believes, is "a living *Today*." We are not enjoying one spark of life in a huge dead waste. We are at a point somewhere in the vast procession of the living, still thinking, still feeling, though invisible and unheard. Our little today, which we often see as the summit of progress, is a tiny fraction of Today itself; but we cannot understand this, cannot enjoy the feeling of liberation it brings us, unless "passing-time falls away from us," unless there is a change in our time-sense.

We need to experience (even if only intermittently, from an indefinable direction, from inside not outside) a sense of *Now*:

No reveries, no conversations, no tracing out of the meaning of phantasies, contain this *now*, which belongs to a higher order of consciousness. The *time-man* in us does not know *now*. He is always preparing something in the future, or busy with what happened in the past. He is always wondering what to do, what to say, what to wear, what to eat, etc. He anticipates; and we, following him, come to the expected moment and lo, he is already elsewhere, planning further ahead. This is *becoming*—where nothing ever *is*. . . . We can only feel *now* by checking this time-man, who thinks of existence in his own way. *Now* enters us with a sense of something greater than passing-time. *Now* contains all time, all the life, and the aeon of the life. *Now* is the sense of higher space. It is not the decisions of the man in time that count here, for they do not spring from *now*. All decisions that belong to the life in time, to success, to business, comfort, are about 'tomorrow'. . . . It is only what is done in *now* that counts, and this is a decision always about *oneself* and *with oneself*, even though its effect may touch other people's lives 'tomorrow'. *Now* is spiritual. It is a state of the spirit, when it is above the stream of time-associations. Spiritual values are not in time, and their growth is not a matter of time. To retain the impress of their truth we must fight with time, with every notion that they belong to time, and that the passage of days will increase them. For then it will be easy for us to think it is *too late*, to make the favourite excuse of *passing-time*. . . . All insight, all revelation, all illumination, all love, all that is genuine, all that is real, lies in *now*—and in the attempt to create *now* we approach the inner precincts, the holiest part of life. For in time all things are seeking completion, but in *now* all things are complete. . . . We must understand that what we call the present moment is not *now*, for the present moment is on the horizontal line of time, and *now* is vertical to this and incommensurable with it. . . . If we could awaken, if we could ascend in the scale of reality concealed within us, we would understand the meaning of the 'future' world. *Our true future is our own growth in now, not in the tomorrow of passing-time.*

In fairness to Nicoll, I must add here that the quotations above, taken from the penultimate chapter of his book, refer to ideas he has expounded at some length in earlier chapters. If in these passages he seems vague and given to easy generalizing, then they and I are doing him an injustice. The whole book should be read.

He has a whole chapter on "Recurrence in the Same Time," but nowhere in it is he as certain and emphatic as Ouspensky was on this subject. He quotes Ouspensky without expressing disagreement or even doubt, but he seems half-hearted, perhaps dubious, about the idea of our repeating our lives over and over in passing time. This does not mean that he does not accept some kind of repetition. Indeed, he insists upon it:

It is difficult to reconcile oneself to the view that a single life determines our lot. We seem to come toward the end of our life just when we begin to get some insight. The illusion of passing-time makes us think that we cannot change the past and that it is not worth while trying to change anything now. We may just begin to realise that we were never taught anything about how we had to live life or about what we really had to do. We probably thought that education taught us how to live, and after a long period of perplexity began to realise that we had to find out things for ourselves. Then it may seem too late. There does not seem to be time for anything and we may easily give up trying to think. The whole impression of the life is a confused one. We do not think we are only *beginning* something, but merely coming to the end of something. If we believe in a judgment and a hereafter, if we believe that our final lot is determined by this single confused life we have led, the idea seems so inadequate that it tends to make us merely shrug our shoulders and turn aside from all such thoughts. It is surely here that the idea of the repetition of the life *necessarily* comes in. . . . Can we not realise that we must have a *future*—that our whole being is constructed to have a *future*. Without some form of the idea of future how can we understand life, how can we possibly interpret it, save in the most negative way? And why should we confine our idea of the future only to a disembodied state which many of us find difficulty in accepting? Remember that we do not live only in this little visible moment but in a WORLD extended in every direction, visible and invisible.

Living Time was published in 1952, and Nicoll died the following year, so he had put the book together more or less toward the end of his life. So had H. G. Wells when he wrote *Mind at the End of its Tether*, a little volume of concentrated despair. As I know from his talk, Wells dismissed the kind of thought we find in Nicoll, without ever closely considering it; but his last book is a dark little dead end and there is liberation, there is light, there is hope for us (if we will get rid of our illusions and wrestle with our vanity and folly) in Nicoll's *Living Time*.

What is revealed in it by way of his particular psychological and philosophical approach is rewarding; his account of our slavery to the idea of passing time is of great value; but while he insists upon repetition, saluting Ouspensky with just a suggestion of doubt, he makes little or no attempt (perhaps because he was a sensible man who had not forgotten his early training) to come close to and then examine the multi-dimensional Time in which he certainly believed.

4.

Like Nicoll, Mr. J. G. Bennett was one of the earliest students of Gurdjieff's system, the Work, and later, like Nicoll, taught it to his own groups. But whereas Nicoll had a medical and psychological training, Mr. Bennett is a mathematician and has been a director of industrial research. With two mathematical collaborators, he has produced a Royal Society paper on *Unified Field Theory in a curvature-free five-dimensional manifold*, which I possess but can only stare at as if I were an Eskimo. (Indeed, there are probably Eskimos now who can begin to understand it better than I can.) Not long after the war he published two books for the general reader—*The Crisis in Human Affairs* and *What Are We Living For?*—which are vigorous and forthright denunciations of various contemporary attitudes of mind, examined in the light of the Work.

And as I write this, he has published the first two volumes of an immensely ambitious opus, *The Dramatic Universe*, being no less than an attempt to bring "all scientific knowledge within the scope of one comprehensive theory of existence." These books, bristling with unfamiliar terms and demanding in places a knowledge of mathematics, are close, hard reading, which I undertook in search of his conclusions about Time. But Mr. Bennett has contributed to *Systematics* a piece about Time, from which he kindly allowed me to abstract the following. In replying to the question "What is Time?," he begins:

There is no simple answer. Our experience of temporality is complex and it varies from one situation

Left, P. D. Ouspensky (1878–1947), at the age of 60. Below, a design made in 1923 for the program of the Institute, which Gurdjieff established at Fontainebleau in 1922.

to another. The variations are so important that we should not speak of time in the singular, but distinguish different times and different kinds of time.

Careful observation shows that time experience comprises three main characters that can vary independently. When such characters can be measured and expressed in numbers they are called parameters and a system with independent parameters is called n-dimensional. We can therefore say that time is three-dimensional. As there are also three dimensions of space, there are in all six dimensions in terms of which measurable, that is physical, events can be described. My co-workers and I have shown that a six-dimensional geometry will serve to describe not only the movements of bodies, but also other physical situations such as atoms and quanta, and the properties of matter generally.

This raises the question whether the three kinds of time apply to other situations such as those of life, consciousness and human free-will. To answer this, we must see what the three kinds of time mean in terms of experience. I shall describe them without attempting to prove that the descriptions are either adequate or exhaustive. I have adopted three names for the different forms of time on the principle that the best way of avoiding confusion is to call different things by different names.

TIME. We experience events as successive. This gives rise to the sense of 'before and after'. The present 'exists' and the past and future do not 'exist' or at any rate do so in a different way from the present. If there were no other time but this, existence would be whittled away into an elusive 'present' that is gone as soon as we reach it.

ETERNITY. We are aware of persistence. Without persistence there could be no change—only the meaningless present. Moreover, there is in every situation a potential for a variety of actualizations. Potential does not come and go as does the actual moment. Pure potential is eternal and imperishable. I have called eternity the 'storehouse of potentialities'. This means that there can be many lines of successive time simultaneously present in eternity. Our experience of changes of consciousness gives us a direct confirmation of alternative times. Furthermore, in all living things there is a persisting pattern that directs their development and regulates their lives. It is impossible to make sense of this self-regulating property of life within the limitations of successive time.

HYPARXIS. The simplest approach to the third kind of time is to consider the requirements of freewill and with it of ethics and responsibility. Successive time does not allow choice. Eternity presents us with the choice, but gives us no room to make it. A third degree of freedom is needed to pass from one line of time to another. This leads to the notion of a third kind of time connected in some way with the power to connect or to disconnect potential and actual. To understand fully the importance of the third kind of time, that I have called hyparxis, we must observe that being itself has graduations. We ourselves can be aware of states when we are wholly controlled by causal influences and other states when we can, not only entertain purposes, but deliberate and choose our actions with the aim of realizing them. I call this variable factor the 'ableness-to-be' present in different beings. It can be traced throughout all levels of existence from atoms through the simplest living forms up to man and it is this factor that entitles us to look beyond man to the attainment of superhuman levels. Without this factor everything would be compelled to remain wholly determined by its own eternal pattern.

The three kinds of time are strictly quantitative—that is capable of being measured and expressed in numbers—only in the physical world. They change from quantity as we mount the scale of existence. In terms of our most intimate experiences, even successive time is not measurable. We can travel in eternity: not in our physical bodies but in our consciousness. We can move in hyparxis by an act of will. But although will and consciousness cannot be measured, they are elements of our experience no less real than sensations of sight and touch by which we know the physical world. . . .

My guess is that Mr. Bennett would not have reached these valuable conclusions if he had not been so long acquainted with the Work, which offered him both a background and a starting-off place. Yet he is not simply giving us Ouspensky with a different terminology. After leaving that starting-off place, he has made full use, on the way to these conclusions, of his own knowledge and expertise. And I for one am grateful to him.

5.

What did the old Master of the Work, Gurdjieff himself, say about Time when it was not masquerading as the "Merciless Heropass"? Only one thing that has come my way. He said "Time is the unique subjective." I do not propose here to explore the depths of that observation. But we do seem to discover that our experience, on any level, is somehow conditioned by something we have to bring to that experience—something not in it but in us. And that something always begins to look like Time. So perhaps Gurdjieff was right.

12 One Man and Time

12 One Man and Time

I.

The "one man" in the preceding chapter heading is of course myself. Who else could it be? And this is not egoism. Let me quote what I said in the Introduction:

But what is more important is that toward the end of this account of Man and Time any attempt at an objective manner would be impossible. The material offered will be itself deeply subjective, belonging to one man's inner world of thoughts, feelings, intuitive ideas, vague impressions, belonging in fact to my own personal encounter with Time. This seems to me the only possible conclusion to such a book. . . .

So far I have done what I could to follow and to consider Time in the outer world—though it would be more accurate to say that my subject really consisted of Time effects. But in the end Time must be tracked down in the inner world, and the only inner world I really know is my own. I repeat, then, this is not egoism. To relate myself to Time is the only sensible and honest way in which to end this book. But I do it in the hope that what I discover in my mind many other people will discover in theirs. It is one of the peculiarities of Time that it is intensely private and yet also widely shared. We could put

We all share the time of watches and clocks; yet the passing of time is an individual thing, different in each situation. For the little boy or the hurried commuter, time may seem to move painfully slowly; for the vacationers in their deck chairs or the embracing couple, time is perhaps almost standing still.

it like this: that superficially, in the world of
clocks and watches and appointments, we share
Time; then, on a deeper level, it seems intensely
private; and then, on a still deeper level, perhaps we begin to share it again, in ways we
cannot yet fully understand.

However, before I begin to relate myself to
Time, there is a moral challenge that must be met,
an ethical obstacle that must be cleared away.

2.

This challenge can be heard in a pronouncement
by a distinguished colleague of mine, who often
expresses scientific opinion. He mistrusted and
disliked theories of Time, he declared, because
they appeared to him "a mode of denying the
seriousness of the moment." And here he seems
to me wrong—and dangerously wrong. To
begin with, the picture he has in the back of his
mind—of so many Time theorists dreamily ignoring the passing of the hours—is quite false.
The Time theorists I have known have been
exceptionally energetic and active-minded persons. They are high above and not below the
average. Nor is it difficult to see why this should
be so. It is their Time theories that encourage
them to appreciate "the seriousness of the
moment." It is their rejection of the conventional idea of Time that helps to give them
energy and to keep them active-minded.

Unlike the majority of people nowadays, the
Time theorist does not believe that moments
flash into our consciousness and then vanish for
ever. These moments make up our lives, and it is
possible, indeed probable, that we do not
deposit what is left of our lives into the grave, all
consciousness leaving us for ever during the last
moment. If we are not hurrying toward oblivion,
if we are shaping a self that will survive death in
some form or other, then our existence in passing
time, moment by moment, does not become less
important but more important. It is the very
idea of multiple Time, with each moment

Left, *The Triumph of Death,*
by Pieter Brueghel (*c.* 1520–69);
and above, *Death and the Maiden,*
by Hans Baldung (*c.* 1476–1545).
The inevitability of death may
haunt all men at some time, but
it is perhaps especially fearful
to those who believe themselves
wholly contained in passing time.

279

existing not only in length but in depth, that brings with it seriousness and a sense of responsibility. And I for one wish to Heaven I had lived with this thought every day of my life.

Nobody can deny that this present age has some extremely ugly features. It has offered us a wide range of nastiness and horror, from dictators ordering millions of innocents into starvation labor camps or gas ovens to teenage city lads amusing themselves kicking old men to death. After despoiling and half-ruining the planet, it has had two world wars and now has plans for a third that would be a man-made Doomsday. Its techniques of mass persuasion are turning men into sheep. And 20th-century urban man is the greatest time waster and time killer this earth has ever known. And in all this global corruption, Time theorists have played no part at all.

But this is not true of the ordinary Time view, the feeling that in the end all our moments add up to nothing, the conviction that our existence is meaningless. It is these that deny "the seriousness of the moment." The moment does not matter because it is only another little step toward final oblivion. No longer is there a Heaven to be won or lost, a Hell to be condemned to; there is nothing. It is all *a tale told by an idiot*. Time is hustling us along to the big sleep.

But what if Time is not as simple as most people now imagine it to be? What if some of the theorists are right? The prospect changes at once, though not all its shades are softer, not all its coloring more harmonious. Suppose we are sentenced not to death, to sleep and forgetting, but to life, to keep on living with ourselves, with what we have been, what we have done? Suppose when we die we rid ourselves only of the world's time, tomorrow's date, but do not escape from our own time and what we have made of it? Feeling sure of oblivion, the suicides put pistols to their heads or ampoules of cyanide between their teeth, to obliterate all consequences of their wickedness or folly; but what if they should find themselves still existing, now without benefit of bullet or poison, and with every consequence still to be faced? What if a man goes yawning on, losing all curiosity and zest and feeling, just

Mummies crumbling away in a catacomb at Palermo, Sicily. Just as death obliterates all life's outward forms, so many of us believe that death—the end of our allotment of passing time—brings oblivion to all consciousness. But such a belief implies that life is meaningless.

waiting to go to sleep for ever, and then discovers after death that he is still awake, still conscious, in a boredom more gigantic and stifling than ever? What if the people who have filled their lives with suspicion and hatred leave only their bodies in the churchyard, their minds then going wandering through a hell they have created for themselves?

Such things may happen or they may not: Nobody knows. As a Time theorist I think the odds are in favor of something of the sort happening, just because I believe that our consciousness does not entirely exist within passing time and that death does not bring complete oblivion. But the point I wish to make here is that it is utterly unreasonable to accuse Time theories of "denying the seriousness of the moment." They not only do not deny that seriousness, they sharply increase our sense of it: They give the moment depth and significance.

One of my B.B.C. correspondents—the only one out of a thousand or more—took a Freudian view of my concern with Time. (He called it Freudian himself.) He said it was born of an

unconscious fear of death. Well, I have never claimed to be an heroic figure, and I certainly share the general apprehension of physical catastrophes—high bridges collapsing, airplanes bursting into flames, head-on crashes of cars, and the rest. On the other hand, having been close to death on several occasions, I have experienced that extraordinary sense of detachment, with events going into slow motion, which seems to suggest that part of our being exists in another kind of Time.

And I certainly do not share the notion, common thoughout a large area of our civilization, that it is ill-mannered and morbid to mention death at all. I feel that many people now are haunted by a curious combined fear of life-and-death, secretly wishing both of them to be scaled down and drained of all color. As Nicoll writes in *Living Time*:

The difficulty is that people do not want to think, to arouse themselves. Even when we are discontented with life we do not want to make the effort of thinking or finding new outlooks. A man wants momentary enjoyment; he does not want to be disturbed; he

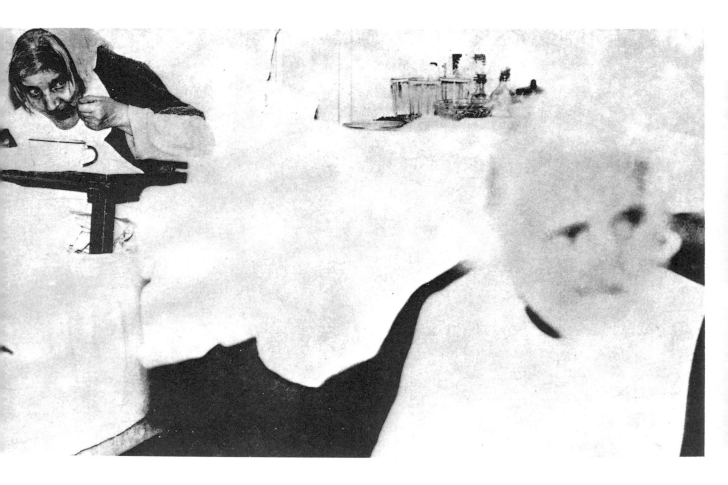

prefers to cling to the opinions he has, and to make everything as easy as possible for himself. . . . Most of us look for rest at death or annihilation. How often does the physician hear the dying say: I want only rest, oblivion—even among those who have appeared to have held the strongest religious opinions. . . .

This is only to be expected, of course, when a man is weary of struggling against pain and weakness, fighting a battle he feels he must lose. But often long before the last sickbed is reached, there is this longing for rest and oblivion. Once youth has gone, people too often find a mechanical existence, with all its meaningless pursuits, more trouble than it is worth. This helps to explain their irritation, their angry comments, when any new theory of Time comes their way. Time the great annihilator must not be challenged. They feel it must be left alone to do its work on them. These people do not want any more life; they feel they have had more than enough already.

I hope the reader will now agree with me that whatever Time theories may be, they are most certainly not "a mode of denying the seriousness of the moment." A wide reading of books and correspondence has convinced me that Time theorists are just as likely to disagree as to agree with one another. But I think they share two convictions. One is that the moment is more serious than most people imagine it to be. The other is that the conventional idea of Time, limiting and enslaving us to passing time, is not only wrong but evil in its power to corrupt us.

Once the mind is freed it likes to strike out for itself, so that I know even now that I am not likely to impose upon many readers my own particular view of Time, just because they will soon have views of their own; but if I have helped to liberate them from this bad idea, still dominating our age, I shall not have written in vain. But, in order to approach my own particular view of time, I must offer a little personal history.

3.

Even as a child I could never understand why certain things that were important to me appeared to older people to be nothing. My dreams were nothing. What I "made up" to delight or terrify myself was nothing. Certain queer feelings, coming out of the blue, were nothing. I can remember, though it must be all of 65 years ago, sitting in the sun on a tiny hillock at the back of our house, and feeling, not lightly but to the very depths of my being, that I was close to some secret about a wonderful treasure, which had no size, no shape, no substance, but all the same was somewhere just behind the sunlight and the buttercups and daisies and the grass and the warm earth. And this too, it seemed, was nothing. I was surrounded and often enchanted, it appeared, by nothings.

And it is true that as I grew up these nothings, once so filled with wonder and terror and joy, did fade and retreat, though never entirely

Not all of us fear the approach of death. The very old and sick may welcome it, as may others for other reasons. Such people seldom want to face questions about eternity ; it is oblivion they seek.

leaving my mind. What happened, of course, was that I had to submit my mind to the pressure of dominating ideas; I was taught to look at life properly, on my way up to becoming a level-headed fellow.

This looking at life properly, with no nonsense about you, and becoming a level-headed fellow, might be compared to attendance at a rather strange movie theatre. In there you are told to concentrate entirely upon the images shown on the screen. These are your world, your life. What is not shown on the screen—yet once again—is nothing. But you cannot help feeling that there is something else, not on the screen. Perhaps you hear a voice that is not coming from there and is much closer to your ear. You seem to catch a glimpse of a face that is not a screen image. There are whispers and movements in the dark. Apparently there is a life all around you, not like the clear and ordered imagery of the screen—a life fragmentary, mysterious, only to be guessed at, but somehow suggesting a fullness and richness of living not to be found in the existence of the lighted images.

Indeed, this screen existence is beginning to seem repetitive and tedious; but one of its hollow-brass voices, probably coming from a machine, tells you not to be impatient; says that

you have only to wait, taking care not to addle your wits with nothings, and soon what the screen will show you will be wonderful. But if you listen hard, another voice, low-pitched, quiet, so close that it might be inside your head, whispers that what you are being told with such authority and complacency is nonsense, that the life around you in front of the screen is real and enduring, and that your nothings have always been *something*.

And then you must make a choice: You can become a sound level-headed fellow, still yawning at the screen and asking when it will all be wonderful, or a bit of a crackpot who somehow keeps on being cheerful and interested.

All that of course would involve other matters besides Time. But Time will do very well as a test. Reduce it to a single line, then almost everything that seems to add richness, depth, and meaning to your mind and to your life becomes a nothing. Or if not a nothing, then a something that a physician or a psychiatrist can rub out, to bring you into the standard pattern of acceptance, boredom, frustration. Now long before I had considered any evidence for precognition and the rest, long before I had read anything about multi-dimensional Time, I had felt a strong resistance to the idea of our being

entirely contained by passing time. It was like being forced into a mental strait-jacket. To accept the idea was to reject thoughts and feelings that, however vague they might be, seemed to light up and liberate the mind. The "facts" might favor passing time, might dismiss any appeal from it; but then I did not really believe in these "facts," though in truth I had never carefully considered them.

What I did believe is that we knew nothing for certain, and that it was better to live between gay tapestries of half-beliefs and fancies than between the iron walls of "facts" that might not even be true. I preferred indulging my fancy to the risk of finally disinheriting my imagination. So I half-believed in a good deal of nonsense, and I still have some notebooks of my teens to prove it. All this was before the First War, when I was a junior clerk in the Yorkshire wool trade. Some of my more fantastic half-beliefs, I suspect, were

Scenes of trench warfare in World War I—the kind of experience through which men can come to realize that the rational "facts" they take for granted in ordinary life may not be so clear-cut after all.

there to defy the wool trade. But then I left it, in 1914, to join the army as an infantryman.

Life in the infantry on the Western Front was capable of squeezing and hammering all the nonsense and fancifulness out of a man. If living hard and dangerously tests a man's notions, then mine were severely tested. So what remained was something quite different from those prewar half-beliefs and fancywork. There was, for example, the feeling (without any theory attached to it) that on some occasions we slipped out of passing time, became detached observers of our fortunes, with death approaching in slow motion, as if we were in some other time. Then again, there were those men, lively gossipers and wags, who became subdued and thoughtful some hours before the sniper's bullet found them or the shell tore their bodies to bleeding shreds— as if they had been watching, throughout a whole morning, death pointing a finger at them across No Man's Land. These men's familiar moods changed completely, we might say, because at the back of their minds a *Now* had opened hugely so that it was already being darkened by an event that was to take place, in passing time, some hours afterward.

And this does not contradict and cancel out that sudden detachment at the high point of

danger that others of us felt, as if we looked on from some other time. True, our *Now* narrowed to the finest possible point, and then opened out, to set things seemingly in slow motion, into another time. The difference—and a difference that perhaps cannot be fully explored by the human intellect—was that these men were going to die and we were going to live.

In this strange region, far removed from ordinary routine existence and its prevailing ideas, a region of hard living and danger and death, I noticed then, though without understanding them, that Time played many tricks. It refused to be true to form. It did not adapt itself to the "facts." And often since then I have noticed that men who have long learned and taught in some such hard school, where you may put up your hand once too often and bring the roof down, are generally averse to positivist certainties about what is fact and what is illusion. Nearly always these men believe we know less than we think we do, and do not care if more advanced thinkers consider them credulous.

Perhaps they are; but then again perhaps in jungles and deserts, on high seas and battlefields, they have seen the certainties fade, the facts turn awkward, and Time play its tricks.

When I returned from the war, and then during the earlier 1920s, I renewed and eagerly enlarged my acquaintance with literature and the arts. With the new novelists—soon to take his place in my mind at the head of them—came Proust, of all novelists the one most deeply concerned with Time, who could declare that the creative part of his being "found itself in the only setting in which it could exist and enjoy the essence of things, that is, outside Time." (By which of course he meant passing time, the fourth dimension, Time One.) And then I began to discover that the contemplation of certain pictures, if sufficiently rapt, seemed to bring about in me a curious Time-shift, so that I appeared to myself to stand and stare at them in some timeless region. This was all the more curious because in those days, for various personal reasons, I felt often desperately hard-pressed, working very long hours to pay my way.

Moreover, because I was by temperament eager and impatient, then and later I was an easy victim of passing time, jumping in imagina-

Passing time does not dominate all 20th-century lives. To the very young, leisurely summer days (left, Monet's *Les Coquelicots*) may seem to be outside Time. Equally, when life is nearly over, Time may have no meaning: Slow strolls can be taken, dogs greeted, clocks and calendars ignored.

tion from one promising occasion to the next (often expecting so much that I robbed myself of any satisfaction) and mentally shoveling away as if they were muck the days between—a bad habit from which I have freed myself only during the last 10 years. So perhaps I denounce the domination of passing time with all the more heat because I have suffered from it myself.

Nevertheless, and in spite of the double pressure of circumstance and my own temperament, even in those years certain works of art seemed to have the power of taking me out of passing time into a spell of timeless entrancement. I am anxious not to exaggerate either the frequency or the force of this power. The entrancement was there, however, bringing with it a *different taste of Time*. Moreover, this could happen outside any appreciation of art. Ordinary life sometimes had this power, as I noted years ago in a chapter of autobiography. For want of a better term I called these particular experiences, even though they had no direct connection with art, "the aesthetic feeling," not because they were associated with beauty, though they were not without beauty, but because they were obviously dissociated from the limitations and pressures and self-centeredness of ordinary existence, and I saw the world as an artist must often see it. This experience, I imagine, cannot be uncommon.

Here is an example. Though I have now forgotten in what city I was, I remember coming to a halt outside a fine large fish shop. As I stared at the scales and fins and the round eyes, looking indignant even in death, I lost myself and all

A child's day, with sand and sea and sun, may seem to happen in an entranced world—but it is a world that need not be entirely lost to adults. Moments when such experience is recaptured seem to bring "a different taste of Time."

sense of passing time in a vision of fishiness itself, of all the shores and seas of the world, of the mysterious depths and wonder of oceanic life. This vision was not in any way related to myself: My ego was lost in it. And real poets, I suppose, must be always enjoying such selfless and timeless visions. They came to me only rarely: It might be from the sight of something, like those fishes gleaming on the marble, or after I had heard somebody merely say "France" or "Italy," or from simply reading the words "eighteenth century"; but they brought me at once a feeling for the immense variety, richness, and wonder of life on this earth. This feeling was deep and joyful but did not belong to ecstatic mystical experience, though it was nearly as far removed from the flat and stale acceptance of everything we find among so many people now.

I felt there was some sort of Time shift, even in those days. Compelled to explain what I felt, I might have said then that I seemed to pass from one kind of experience with a certain temporal effect to another and quite different kind of experience with another and quite different temporal effect. What I did not understand then was that there might be more than states of mind involved here, that there really are different orders of times within Time itself, that our consciousness dwells among many dimensions.

As I said earlier, Dunne's *Experiment with Time* came out in 1927, and as I had it to review I read it very carefully. But for some reason or other (possibly because I had decided to write at least two long novels), though I was fascinated by Dunne's book it did not arouse in me an immediate interest in the Time problem. What it did do—and this happened to many, perhaps most, of its readers—was to make me increasingly conscious of my dream life. Though I never in fact adopted the notebook method he urged upon his readers, I did begin to remember, sometimes in detail, more and more of my dreams. This was fortunate because as I entered the middle 1930s I began to have some extraordinary powerful dreams, a few of which I described in print.

These middle 1930s were my own early 40s, a climacteric period when, as Jung has so often pointed out, a man must face squarely the second half of his life, which should have an entirely different character from that of the first half. I was then beginning to read Jung, whose theory of the collective unconscious seemed to me—as it still seems to me—one of the great liberating ideas of this age. And it was not long before I was dividing such close attention as I could spare from my work—then making great demands upon me, especially in the theatre—between Jung and the Time problem, always regretting that this division was necessary. If Jung had given as much of himself to Time as he did to alchemy, I am certain that this book, if still necessary at all, would have been much easier to write.

Now I must say something about my plays, not because some of them have been called "Time plays" but because they involved me in a Time shift very different from the one I had known earlier. I shall be as brief as possible but some explanation is necessary here. During this period I wrote both comedies and serious plays. Contrary to general opinion, it was the comedies that took me far longer to write and demanded far more conscious effort. After some preliminary pondering over their characters, I wrote almost all the more important serious plays at a furious speed, without making any scenario or elaborate notes, and with little or no conscious effort. For example, *Dangerous Corner*, *Time and the Conways*, *An Inspector Calls*, *The Linden Tree*, were written in 10 days at the most, probably within a week. Their most difficult acts, those bristling with technical problems, demanded no more effort than the others (and later were rehearsed and played with only a few words changed); and I might have been writing letters to friends.

The sheer speed of composition made any conscious appreciation and solution of the technical problems involved quite impossible. Looking at them afterward, I felt like a man watching himself run at a headlong pace across a mine field. And experience played little or no part in all this: *Dangerous Corner* was my very first play. Moreover, between these headlong pieces, I was quite capable, working slowly and making an effort, of writing very faulty or even abortive plays. Yet out of the four pieces of

furious creation named above, the first three have been played more often and in more different countries than any of the 20-odd other plays I have written; and the fourth, *The Linden Tree*, less likely to be understood abroad, was an instantaneous success with the English audience for which it was written.

In spite of this astounding speed of composition, it did not occur to me then that any Time element was involved. With Jung's theories in mind, I felt that the hard work in this apparently effortless playwriting had somehow been done in and by the unconscious, which had then broken through and taken charge and used my conscious mind simply as a transcribing instrument. So it did not occur to me that there was any Time element in this almost magical creation—for however modest its results may have been in terms of world drama, it was almost magical to me.

But now I see that we cannot rule out Time, which has its own relation to the unconscious. We know that on one level the unconscious is capable of keeping an eye on chronological time for us, waking us if necessary at any hour we choose. But this is not its own time. It refuses to accept, when it is about its own business and not acting as an alarm clock, our whole idea of temporal succession. Its time order is not ours, as Jung himself pointed out to me, some years later than this playwriting period, when I had some talks with him. He himself, I felt then, wanted to keep clear of the Time problem, though afterward he may be said to have challenged conventional and positivist opinion with his curious, fascinating, if rather obscure essay on *Synchronicity*. I call this essay a challenge because the "acausal connecting principle" it suggests (after showing us some astonishing groups of "coincidences") can hardly be compressed within a uni-dimensional linear time.

There were then these two different kinds of experience, only alike in appearing to suggest some sort of Time-shift and in releasing the mind from an egocentric relation with passing time. In all other respects they were so different that it is hard to believe that any one order, relating either to Time or to states of consciousness, could contain them both. Whether it came at moments

of great danger, in contemplating works of art, or with "the aesthetic feeling" about certain aspects of life, the first kind of experience put things into slow motion, detached my consciousness from passing time, and transformed me while it lasted into an almost selfless observer, existing outside any sphere of action.

The second kind of experience did not withdraw me from action but flung me into it, did not turn me into a detached observer but into a creator working like a man possessed, lending me energy and imagination and a creative will. But I write "me" for convenience; actually in this experience there is an absence of any feeling of self; and if some spirit, not a member of the Society of Authors, claimed this writing I would feel no resentment.

We have here then two fundamentally different kinds of experience. They belong to different states of consciousness. But there is also a Time element here. They are alike in appearing to be outside passing time. But my mind seemed to make its escape from passing time, so to speak, in two quite different directions. In one there was room for contemplative inaction; in the other there were possibilities of the most rapid and decisive action. (This would not be true if I were not by choice and profession a writer, to whom writing is essentially "action.") And for my part—and, after all, they were *my* experiences, and any critical Dr. Brown or dubious Professor Smith may never have known them— I cannot detach their difference and their value from their Time character.

Time seems to divide itself into times here. There is one for passing time; there is another for the first kind of experience, the contemplative slower-up; there is another for the second kind of experience, the purposeful, imaginative, creative speeder-up: three times. So can it be true to say

Two drawings from *Opium* (Jean Cocteau's diary of his cure from addiction). A drugged mind and perception is in a state parallel to that of dreaming: "Observer 2," we might say, has taken over; normal vision (and normal Time) has been set aside.

that nothing in our actual experience suggests—if we want to be geometrical about it—that there might be three dimensions of Time? I say it cannot be true. I will agree that no exact analysis may be possible, that no sharp lines can be drawn, that all except one's innermost feeling is blurred and shadowy, that the relation between consciousness and the unconscious may complicate the issue; but I cannot escape the feeling that Time, itself so blurred and vague and elusive, divides itself into three to match these different modes of consciousness. We are at least entitled to say that it is *as if* there are three kinds of time. (And this is hardly an impudent claim when Time itself, on close examination, can be turned into an *as if*, even though we have in this age transformed it into a ball-and-chain to keep the spirit a prisoner.)

4.

At the risk of appearing to put myself too close to Dunne, I propose to call these three times—time One, time Two, and time Three. To follow some theorists and call time Two "eternity," a term rich in associations, would only mislead and confuse many readers. Again, though it would not be difficult to invent a name for time Three—like Mr. Bennett's "Hyparxis"—some of us are easily repelled by unfamiliar terms. So let us make do with times One, Two, and Three, remembering that we live in all three of them at once, though we may not enjoy, so to speak, equal portions of them.

As visible creatures of earth we are ruled by time One. We are born into it, grow up and grow old in it, and die in it. Our brains have developed through eons into marvelous instruments of time-One attention. Not only do they bring to our notice almost everything we feel we ought to know, but they are able to exclude what might be bewildering and unhelpful. When drugs interfere with their chemistry, some of their inhibiting processes do not work, and then we might see a chair as a Van Gogh might see it, not as a furniture salesman and a customer might see it. (The Time-shift in drug experiences is well attested; they free consciousness from its age-old concentration upon time One.)

Our relation through the brain with time One tends to be practical and economic, good for our matter-handling business, which helps to explain why we are now great time-One people and mostly try not to believe in anything else. If this limited belief were imposed on people as a dogma, as it easily might be in totalitarian or severely conformist societies, it is not merely fanciful to suggest that men might become automata, ruled by better machines than themselves.

There are signs, however, of a reaction against this time-One dogma. Many of them have come my way since I began writing this book. Not all the forms this reaction takes are acceptable. One that is acceptable, concerned with all the extrasensory perception phenomena, ESP at work, which only sheer bigotry can deny now, seems to me outside the scope of this inquiry. But as we have seen already, even though we ignore all manner of examples of premonition and *déjà vu* experiences, there are plenty of authenticated precognitive dreams to prove that our minds cannot be entirely contained within time One. So let us take an early example of the precognitive dream, one connected with an historical event, and see what we can make of it in terms of more times than One.

Three months before Napoleon invaded Russia, the wife of General Toutschkoff had a dream that was repeated a second and then a third time in one night. In this dream she was in an inn she had never seen before, in some town she did not know, and her father came into the room, leading her small son by the hand, and told her in broken tones that her happiness was at an end because her husband had fallen at

Above right, the battle of Borodino, 1812, depicted by the French painter Baron Louis Lejeune (1775–1848)—the battle in which the husband of Countess Toutschkoff died, as she foresaw in a dream. Borodino (on the map, right, of Russian positions at the battle) is a small village about 70 miles from Moscow; it appeared on none of the countess's maps.

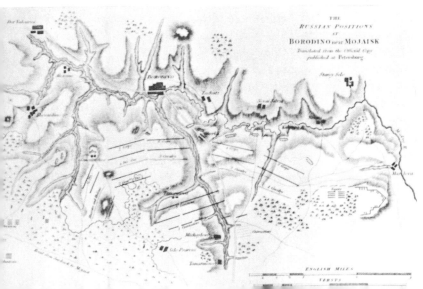

293

Borodino. She awoke in great distress, roused her husband, and asked him where Borodino was. But when they looked for the name on the map, they could not find it. (The battle in fact took its name from an obscure village.) After it was fought, everything happened as in the three dreams: She found herself in the same room in the same inn in the same town, and her father came in with her son and announced her husband's death at Borodino, where he was commanding the army of the reserve.

More or less following Dunne here, we can say that the dreaming self of Countess Toutschkoff, in time Two, revealed what would happen to her in time One. Though a soldier's wife might always be haunted by the fear that her husband might be killed in battle, coincidence must be ruled out of her three dreams because what happened was identical in so many different particulars and that the very name afterward given to the battle was then unknown to her. If no part of her mind could escape from time One, then the whole matter remains inexplicable.

In the dream quoted on pages 222 and 295, a man knocked down a boy with his car (above left). When the same incident occurred in reality a few weeks later (right), the man was able to avoid the boy: Apparently he was able to take action to change the foreseen future.

Another Time order, which we can call time Two, does at least offer us a possible explanation. But what about the historical event, the Battle of Borodino? Are we to assume that before Napoleon's army crossed the Niemen on June 24, the Battle of Borodino on September 7 was already waiting to take its place in history? And if it was not, as we cannot help feeling, if all depended on the calculations of the French and Russian general staffs and the consequences of various minor battles and skirmishes, then where did the dreamer's mind, wandering in time Two, discover this unknown name Borodino? If in our instinctive dislike of the idea of a fixed unalterable future, we declare that early in June 1812,

when the Countess dreamed her three dreams, events that were to take place in September did not exist in any possible shape or form, then how could she dream as she did? And if we accept her precognition, and with it the idea that this was an experience in time Two, impossible in time One, then how do we avoid the fixed future, the Borodino already in its place further along the track of time One?

Before I try to answer those questions, let us return to the dream of the American mother about the visit to the creek and the dead baby (page 225). After describing that dream, I said that if we accepted it, we must also accept one of two things:

We must believe that up to the moment when the mother leaves the baby to go and fetch the soap, the dream is showing her the future, but that her return to find the baby drowned is a dramatization of not unusual maternal anxiety: the dream being therefore part-future, part-fiction. *Or* we must believe that a future containing a dead baby is *changed*, by the mother's action, into a future in which the baby does

not die and lives to become a father himself: so that of two possibilities, one by deliberate intervention has come to be actualized. This leaves us with a future already existing so that it can be discovered by one part of the mind, and with a future that can be shaped by the exercise of our free will. We cannot have both, we shall be told; it must be either one or the other. Possibly, possibly not.

It may be remembered that I offered a similar alternative after an account of a motorist who dreamed that he knocked down a small boy and then, a few weeks afterward, had to swerve and brake violently to avoid hitting a small boy whom he immediately recognized, after getting out of his car, as the child he saw in his dream. I said that this dream could be part prevision, part fiction, or that it could show a possibility never actualized because the driver, forewarned by the dream, was able to act promptly at the right moment, changing the future seen in his dream.

In both instances, then, we are left with a choice: between a dream that only in part reveals the future and a dream entirely con-

295

cerned with the future—but a future that is not fixed and inevitable, that can be changed. Let us consider the latter first. We are asked to accept a future that exists in some form or other, because it can be experienced in a dream, and yet may possibly be changed. Thinking theoretically, we feel inclined to reject at once any such idea of the future. Either the future is an "uncreated nothing" or it is wholly there, waiting for us to experience it. But this is only what we think in terms of Time theory.

In our ordinary thinking, outside theory and well inside practical living, not only do we not reject this idea of a half-made future, consisting of possibilities that may or may not be actualized —that is, becoming part of our and the world's physical history—but we accept it so whole-heartedly that it shapes and colors our thought. When we think of the next 12 months we regard them neither as a blank nothing nor as some inflexible series of events. These extreme alternatives belong to theory, not to our actual practice, in which we do not hesitate to steer a course between them.

It is this intellectually infuriating future, rather like an omelette just before it is ready to be lifted out, that we hold in our minds when we are actually planning our lives and not picking and choosing among Time theories. Certainly an H-bomb missile may arrive and blow us all out of time One, but we are well aware of that, know that it is one of the possibilities that may be actualized. That missile, we may say, is waiting to be sent on its dreadful errand, to mark a high point of human folly; but it has been put together by man's will and action and it can be outlawed and destroyed by man's will and action. The future cannot be nothing, for it contains immediately, among other things, that missile; but neither is it fixed and inevitable, for it can be changed even after it has seemed to reveal itself in dreams and time Two.

It is then not at all outrageous to suggest that these two dreamers saw a future that existed in some shape or other (notice how we like saying, with Wells, "The shape of things to come") and yet could be dramatically changed by a sudden act of will—the picking up of the baby, or the swerving and braking of the car. After all, it is far closer to our common working idea of the future —half-made, half-there, not wholly made or wholly there—than any conclusions of the scientists or the philosophers. We have in these dreams possibilities that were · actualized— namely, the visit to the creek, the encounter with the boy on the road—and possibilities that were prevented from being actualized—namely, the drowning of the baby, the knocking down of the boy. (With the Borodino dream, of course, everything was actualized, probably because an event on that scale, a great battle, is fixed beyond intervention.)

Now in time Two, where the dreams belong, there is no distinction between the possibilities that are actualized and those that are not. The creek and the road are there, and so are the drowned baby and the knocked-down boy. In time One, action can be taken so that what we might call the line of history avoids the drowned baby, the injured boy. But that is all. The alternative possibilities, together with the choice between them, cannot exist in time One. Nor did they exist in time Two. Therefore another time is necessary, time Three.

If we think of this line of material history really as a line, then as one possibility is actualized and others are not, this line cannot move up and down in two dimensions but must curve around in three. So on this theory of the two dreams we can accept Ouspensky's "The three-dimensionality of time is completely analogous to the three-dimensionality of space" and J. G. Bennett's "This leads to the notion of a third kind of time connected in some way with the power to connect or disconnect potential and actual." And certainly, taking the baby away from the creek, swerving and braking to avoid the small boy on the road, may be said to be disconnecting potential and actual.

I refuse to answer questions, however, about these possibilities in time Three, these shapes of possible things to come, which can be seen in time Two and yet, until actualized, are outside our material history in time One. I refuse to answer because I do not know. Nobody should be surprised by this confession. After all I do not know, can only hazard a guess, how the sight of a picture or even the sound of a word could suddenly seem to change the tempo and tone of my existence, apparently taking me out of one time and enclosing me within another. I do not know, and again can only hazard a guess, how I came to write, let us say, the technically complicated Act Two of *Time and the Conways* seemingly without effort and at a headlong pace. All that is within my sure knowledge is that these things happened. But then every other day something happens that a positivist would call a coincidence and that I feel is nothing of the sort, though I cannot explain it. I am old enough now to realize how little I do know.

Let us see now what happens—and here I have little more advance information than the reader has—when we reject the idea that these dreams were entirely concerned with the future, but a future that could be changed. We put in its place the idea that these dreams were, as I said earlier, part-future, part-fiction. The visit to the creek with the baby, in the dream, is a genuine time-Two glimpse of the future, but the tragic episode of the baby's death does not belong to the future but is, as I said, "a dramatization of not unusual maternal anxiety," the kind of thing most women cannot help imagining.

In the same way, that section of road and the small boy on it were there in the future, seen by the dreamer in time Two; but projected into the scene, out of the motorist's constant anxiety about accidents, was the imaginary episode of injuring the boy. In these terms the future remains unchanged; all that happens is that what was imagined did not take place in reality. And at first sight this seems a much neater and more sensible explanation.

After reflection, however, this explanation leaves me feeling dissatisfied. The questions that seem to vanish have merely been swept away like crumbs and ash under a rug. Anxiety dreams are common enough; most of us have wakened to remember vaguely those dream airplanes and trains that will not be caught because taxis break down and luggage is mislaid. But in these two dreams there is no such dim confusion. As in most precognitive dreams, everything is sharp and clear, so vivid that it is easily remembered. And there does not appear to have been any break, any change of quality, between what was previsionary, disclosing the scene and the situation, and what belonged to a dramatized anxiety. The dead baby seems to have been as convincing as the live baby.

However, I do not wish to attack the theory from this position, if only because so much is possible in dreams, those vast private theatres in which we are dramatists, directors, designers, actors, and audience. (Long before I was interested in precognition, before I had read any depth psychology, I could not understand why so many people thought their dreams no more important than their sneezes and yawns. Our dreams are our night life.) I will suppose it to be true that a dream can move smoothly from prevision or precognition, from what is revealed, not created, to this familiar acting out of an anxiety or constant fear. Nevertheless, I am still left feeling dissatisfied.

Certainly we have now dodged the future that can be experienced and yet changed, and those possibilities that may or may not be actualized. It is much easier to say that something was imagined—such as the death of the baby—that did not subsequently take place in reality. It is easier still if we assume, as so many people seem to do, that the imagination and all its creations are nothing. But this will not do, for these are very much *something*, and something of immense importance to our lives. (What if we should finally arrive, a long way beyond death, at a mode of existence that remained nothing but a huge blank unless the imagination went to work on it, creating for it landscapes, cities, dwellings, gardens, arts, and social life?)

Because most children are highly imaginative, it is supposed by some that to reach maturity we ought to leave imagination behind, like the habit of smearing our faces with jam or chocolate. But an adult in whom imagination has withered is mentally lame and lopsided, in danger of turning into a zombie or a murderer. It is the creative imagination that has given our ruthless bloodthirsty species its occasional gleams of nobility, its hope of rising above the muck it spreads.

To discover what we make of imagination I am consulting the nearest *Dictionary of Philosophy*, an unambitious American work, usefully modest in size. It says:

Imagination designates a mental process consisting of: (a) The revival of sense images derived from earlier perceptions (the reproductive imagination), and (b) the combination of these elementary images into new unities (the creative or productive imagination). The creative imagination is of two kinds: (a) the fancy which is relatively spontaneous and un-controlled, and (b) the constructive imagination, exemplified in science, invention and philosophy which is controlled by a dominant plan or purpose.

To which I feel I must reply: (a) that there is here a smell of the cobbler's "Nothing like leather"; (b) that I should like to read the com-ments of Samuel Taylor Coleridge and William Blake on this account of imagination; (c) that the enduring vitality, say, of Don Quixote and Hamlet, apparently the products of uncontrolled fancy, is not easy to explain along these lines; and (d) that if this can be taken as a representative statement, then we know and care very little about imagination. It really belongs to the invisible nothing department.

Because imagination appears to be free of the limitations we know in time One, we think of it as being outside Time. It is there, however, that the nothings begin. We might do better if we thought of it as belonging to a different Time order, to another time. And I have already suggested, after remembering my own experience, that imaginative creation seems to imply not a second time order, contemplative and detached from action, but a third, in which purpose and action are joined together and there seems to be an almost magical release of creative power.

If there is a part of the mind or a state of consciousness that is outside the dominion of time One and time Two but is governed by a time Three, then that is where the creative im-agination has its home and does its work. And it may be that there imagination is not something escaping from reality but *is itself reality*, while the world we construct from our time-One experi-ence is regarded there as something artificial, abstract, thin, and hollow.

Now what belonged to the time-One future in these two dreams, the baby at the creek, the small boy on the road, could only be observed in the wide but badly focused present (compare Dunne) of time Two. This is what the dreamers are attending to, being asleep and incapable of concentrating their attention in time One. Now we are still assuming that the main incidents of the dreams, the death of the deserted baby, the running-down of the small boy, are not part of the future revealed in time Two. These tragic incidents have been neatly introduced into the dreams, perhaps to serve as warnings, by the dreamers hastily dramatizing familiar anxieties. Thus the set and the characters are provided by future experience, but the action comes out of imagination. This is all very curious, but it is what we must accept on this theory of the dreams.

And Time cannot be left out, because we know the future is involved; and if Time is in, then it cannot be uni-dimensional, for we need another and different time, a time Two, to explain how the future came to be revealed. But this will not explain what we have already assumed: the sudden change in the character of the dreams, the switch-over from prevision to imagination, the dramatic intervention of the dreamers' anxieties and fears.

In which time order did imagination come to intervene? Not in time One, which is closed down for the night. Not in time Two, which is concerned with the time One future and is pro-viding the baby at the creek and the small boy on the road. The dramas, the warning strokes of the imagination, which will eventually produce decisive actions in time One, must belong neither to time One nor time Two but to time Three. So, though by a very different route, we arrive again at that third dimension of Time.

Nevertheless, I prefer the other explanation of these dreams, which does not see them as a mixture of prevision and imagination (even though we do not know what imagination is) but as glimpses of a future already shaped but still pliable, yielding—in these instances, though obviously not in many others—to will and action. It is as if ahead of us in time One were shapes, molds, patterns, possibilities, seen as definite

One of William Blake's drawings for his *Jerusalem*. In the center, Los stands between life and death. On the right, the moon and stars preside over the dark entrance to the Serpent Temple (the pathway of experience); on the left, the sun of Time is carried back to eternity.

events in certain of our dreams; and into some of these shapes, molds, patterns, there arrives the material substance that actualizes them, hardening them into world history. There is an idea not unlike this in Blake's *Jerusalem*, in which *Los* (the name is *Sol* reversed) can be taken as the symbolic figure of Time:

> *All things acted on Earth are seen in the bright Sculptures of*
> *Los's Halls, & every Age renews its powers from these Works*
> *With every pathetic story possible to happen from Hate or*
> *Wayward Love; & every sorrow & distress is carved here,*
> *Every Affinity of Parents, Marriages & Friendships are here*
> *In all their various combinations wrought with wondrous Art,*
> *All that can happen to Man in his pilgrimage of seventy years. . . .*

These "Sculptures," as Maurice Nicoll suggested, can be regarded as states of mind from which men cannot free themselves; but they can also be seen as the possibilities, like the Borodino of the Countess's dream, that must be actualized. They are that part of the future that is fixed. But much of it, close to us as individuals, is only half-made, depending for its historical time-One shape and character on a number of personal decisions.

This brings us to the questions I asked at the end of the chapter "The Dreams." There, after pointing out that more than 90 per cent of the precognitive dreams I was sent were concerned with the terrible or the trivial, I asked why the dreaming self so rarely catches a glimpse of that wide middle range of our activities and interests. And now we can reply that within this range there may so often be no determined future, only a confusion of possibilities still to be actualized, waiting to receive their time-One shape and character from will and action. Both the terrible and the trivial are nearly always outside our control—the deaths and disasters because they are too big and fateful, the trivia because they are too small and unimportant. So that it is they in nine cases out of 10, at least, that will be revealed in precognitive dreams because they are there to be revealed. These possibilities are part of the future that will not be changed, either because they are out of reach of our will or beneath its attention and interest.

Along this line we can now approach the FIP (future-influencing-present) effect. Why did my friend Dr. A feel a queer excitement when he received impersonal official reports from Mrs. B, then completely unknown to him? His consciousness in time One knew nothing about her. No dreams from time Two visited him. But in time Three they had already fallen in love and were married. This deep relationship, for some reason I cannot supply, was not a possibility but a certainty, but only in that remoter part of his being—if I may be allowed the phrase—involved with time Three, able to communicate nothing to his time One consciousness but this queer feeling of excitement.

This is no doubt an exceptional instance, but what is not at all rare, at least in my experience, is a state of mind suddenly and inexplicably illuminated or darkened by feelings apparently coming from nowhere and entirely unconcerned with what we are doing, thinking, feeling, in our time One existence. This last rightly demands most of our attention, but we must not make the mistake of assuming that anything not explicable in time One terms is a nothing. It may be a very important something, like the excitement that Dr. A felt, the distant trumpets heralding the most rewarding relationship of his life. We should think a man a fool if he insisted upon meeting all his time-One experience with half-closed eyes and with wax pads over his ears. But we shall not be very much wiser if, to prove too narrow a theory, we try to keep our minds closed to what might be revealed to them in times Two and Three. In this way we could impoverish our experience both on this side of the grave and then beyond it.

I have lately received, from Italy, some material based on the findings and theorizing of a small newish international group of medical psychologists. This restored to me a term much used before the First War, when I first began to read about and discuss such matters, but one I have rarely seen or heard since then: This is the "superconscious." On this theory the ego and its field of consciousness occupies a middle place between the unconscious, personal or collective, and the superconscious, the source of our nobler feelings, intuitions and inspiration, genius, illumination, and ecstasy.

And if we relate this division to our temporal system here, we could say that the ego and its field of consciousness belong to time One, the unconscious to time Two, the superconscious to time Three. But we must remember there are no separate compartments and exact divisions, and that we live, even here and now, in all three times. This still remains true even of those people who deny the possibility of any experience outside time One. However, it may not always be true, for if we insist upon disinheriting ourselves, men may ultimately become time One slaves or automata.

5.

I have said that we go beyond the grave. But in one sense and strictly speaking, we do not. Indeed, it is the idea of a time-One existence persisting after death that has worked so much mischief. It has helped to create some of the dreariest fantasies ever known, with the spirits of departed Red Indians arriving in South London basements to establish communications between this world and the next. It has so repelled many people that they proclaim with passion that long before the doctor has signed his certificate we shall be dead as mutton. The truth is, we are apt to be immodest in both our claims and our denials here: Either we live for ever or perish with our last heartbeat. I think it reasonable to suggest we do neither. We are not demigods and we are not cattle.

We cannot go beyond the grave in time One. When we die we come to the end of our allotment of time One. The brain ceases to supply us with any further information because it stops working, dying with the body that housed it. We have to take our leave of chronological world time. We move out of history: *Thou thy worldly task hast done, Home art gone, and ta'en thy wages.*

But where is home, and what are our wages? In terms of our argument here, we can say that home is our continued existence in time Two and that our wages, there for us to spend in that existence, are our total experience in time One. This has an end just as it had a beginning in time One; we go jogging along that world line until death appears as a terminus; but in time Two we have never been making that journey and, as we have seen, have never been bound by its conditions. There may be some kind of death awaiting us ultimately in time Two, but it is certainly not the familiar time-One death. This we survive because our consciousness has never been contained within time One.

And it is quite beside the point to object that we have no visible evidence to prove that survival. Time Two, in which we do survive, does not work in the visible-evidence department. I cannot take that "aesthetic feeling," a time-Two experience, into a laboratory to be weighed and measured. The most meaningful and the most ecstatic moment I have ever known occurred in a dream, not precognitive but deeply symbolic, a dream that changed my whole outlook; yet I have for it less visible evidence than I have for a slight cold in the head or a broken fingernail.

Our time-Two world, in which we survive, has as its foundation our total experience in time One. Now it cannot be denied, for this has been proved over and over again, that the brain acts as a marvelous recording instrument, storing somehow and somewhere an exact impression of every moment of our lives, something quite different, in its brilliant immediacy, from what

An illumination from a 15th-century French manuscript showing God receiving a dying man's soul. Most people in western Christian society who believe in an eternal life-after-death see it not as a different "kind of time" but as an endless extension of their existence in passing time.

may be recovered by the ordinary memory. (Under hypnosis or the emotional pressure of drastic analysis, middle-aged men and women have been suddenly turned back into children of two, destructive and screaming with rage.) What the relations are between this stupendous brain storehouse and the mind or consciousness and the unconscious, I do not know, and I do not think anybody else knows.

And I am not denying that this endless recording, much of it inaccessible to our day-by-day consciousness and seemingly out of all proportion to our needs, may play a part, not yet understood, in our time — One existence.

But I believe with Dunne that when we have come to an end in time One, we go forward—spatially and geometrically, we may say, at a right angle—in time Two, no longer concentrating our attention on the physical world, but now having in place of it all that accumulation of mental events, all the sensations, feelings, thoughts, left to us from our time-One lives. It must be remembered, however, that we have never been living exclusively in time One, however much we may proclaim that no other time exists. So it might be as well for us hereafter, when we are out of passing time, that we do not think and feel and behave now as if passing time were all we had. And surely this is what most religions, behind their popular melodramas of angels and demons, saviors and devils, heavens and hells, have been trying to teach us.

Because we have never been living exclusively in time One, we are not without clues—that is, if we do not willfully ignore them—as to what might happen when we are no longer in passing time. I have already given some examples of what we might call time-Two and time-Three effects, taken from my own experience. Now here is another. Can we imagine ourselves in a fifth dimension (time Two) obtaining a four-dimensional impression of a fellow creature? Alarmed by these dimensions, about to create a monster, we will probably reply at once that we cannot even imagine such an impression. But unless we have been unlucky in our relationships, I maintain that we are always taking what is in effect a four-dimensional view of the persons nearest and dearest to us. It is in fact impossible to avoid taking this view in a close, deep, and lasting relationship.

We do not see these loved persons entirely in passing time, as three-dimensional cross-sections of their real selves. We habitually see them, somewhat out of focus in passing time, in a curious blur that releases tenderness, not only as they are but also as they might be and as they were, reaching back, if we are parents, to their earliest childhood. The eagerness of lovers—and this is especially true of most women, usually more aware of this four-dimensional effect than most men—to know, to see, what the other was like years before they met, seems part of an instinctive desire to enjoy this deeper impression as soon as possible. It is an attempt, inspired by profound emotion, to reach beyond passing time, to fix the relationship in different and more enduring conditions of experience.

Then dreams, of course, offer us other clues as to what might happen when we are no longer in time One. This does not mean that dreams in general—and now we set apart those rare clear dreams—can be taken as examples of our time Two existence. Allowance must be made, as Dunne pointed out, for the confusion between two times, two outlooks. When we no longer return to time One, the situation is altered. (But there is a Buddhist tradition that a man who can control his dreams while dreaming will control his states of being in the after life.) We shall have to learn how to live in time Two, which might well seem at first an uncontrollable dream world, through which our consciousness wanders like

Goya and the Duchess of Alba, painted by Goya in 1793. At the time, she was 31 and he 47; yet both figures are youthful and attractive, as though Goya had taken a "four-dimensional" view of their relationship and fixed it at its most perfect.

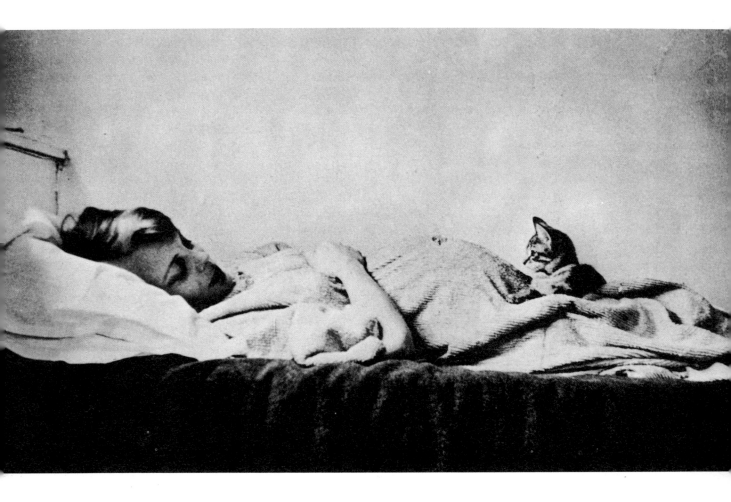

Alice on the other side of the looking glass. Because, as in dreams, we can no longer know certain time intensities, we shall miss the keenest sensual satisfactions but we shall also be beyond the reach of the sharpest pain. On the other hand, again as in dreams, what might be called our emotional landscape may be immensely enlarged and far more highly colored, mountains of wonder and joy rising above sinister depths and chasms of terror.

Any notion that wish fulfillment is at work here, sketching this time-Two and time-Three existence beyond time One in soft pastel shades, can be dismissed at once. We shall not sink into Abraham's or anybody else's bosom. We shall not be little lambs gently carried into the fold. The last flicker of our time-One consciousness will not cancel out for ever our follies and malignities, allowing us the sleep of the just when we have been so unjust. (Because so many people

believe or hope it will, they no longer feel responsible.) It is here, in the world we have made, we really begin "to live with ourselves," and reap between these heavenly heights and hellish depths what we have sown. And we cannot say we have not been warned; we have been warned over and over and over again.

Anybody who can find wish fulfillment here must feel a great deal more complacent about his time-One life that I do about mine. I can see this time-Two existence, with time Three and its fiery creative energies now a new time Two, offering us some very rough going. Courage, imagination, and love, which we praise in a routine fashion more often than we really try to achieve, may be as urgently necessary as air and water and bread are now. So we might be well advised to stock up while we can. (It is what we were always told to do—that is, before we became members of an affluent super-technological

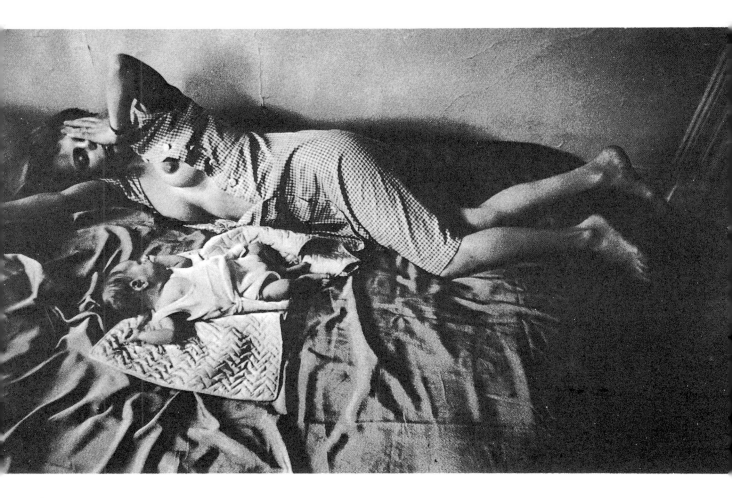

society, giddy with conceit because we might put a man on the moon.)

There is, however, one feature of this time-Two after-life that might seem to suggest a kind of professional wish fulfillment on my part. For I am by profession and temperament a dramatist, and I detest ill-contrived and under-rehearsed scenes. But my time-One life can show me—and indeed *will* show me when it turns into my time-Two world—far too many of such scenes. So I welcome the chance not of simply re-living them, though that may have to be done, but of *beginning to put them right*. For this I believe, quite apart from any professional and temperamental bias, we shall have an opportunity of doing, if we can work with others, whose lines cross ours in this time-Two world, in trust and love. We may have the choice—and this involves no intervention from higher levels of being; the choice could be entirely ours—between building

A mother's love for her child transcends passing time, seeing the child not only as it is but as all that it might become.

305

a self-glorifying palace out of our time-One material, until we wall ourselves into a hell of loneliness and desolation, and trying to create in trust and love, at what we might call the crossroads of our respective world lines, a new and more rewarding life. We can begin to do this here and now. But we can do it better there, on the other side of that first but not last curtain of Time.

Even while we are on this side of that curtain, however, is there not always a something else that can never be fitted into the time-One pattern? I am not thinking now about precognitive dreams, premonitions, and the like, nor about the contemplative-aesthetic and the imaginative-creative experiences I described earlier. This something else is impossible to prove and is hard to capture in words. It can come at high moments of love and, though rarely, at a meeting of friends; it can transform some sudden glimpse of a landscape into an irradiated sign; it haunts some music for us; its light and its strange shadows fall on certain scenes in drama and fiction. Always it adds depth to life, suggests an ampler Time, opens a new dimension.

It never stays long, at least for most of us, but if, fixed in our attention to time One, we cease to be conscious of this something else, this bonus from the unknown, really arriving from another mode of Time, we begin to feel stale and weary. We are not only not preparing ourselves for existence in the next world, we are beginning to lose interest even in this one, for the scene is flat and its colors are fading. To die is not to close our eyes when we come to the end of our time One: it is to choose to live in too few dimensions.

6.

My personal belief, then, is that our lives are not contained within passing time, a single track along which we hurry to oblivion. We exist in more than one dimension of Time. Ourselves in times Two and Three cannot, vanish into the grave; they are already beyond it even now. We may not be immortal beings—I do not think we are or should want to be—but we are something better than creatures carried on that single time track to the slaughter-house. We have a larger

portion of Time—and more and stranger adventures with it—than conventional or positivist thought allows.

But it is still a portion; we have not unlimited Time, though what the limits will be in time Two and time Three, I do not know. Nor of course do I know what happens. I suspect, however, that in time Two we begin by being more essentially ourselves than in time One but end by being less ourselves, personality as we know it vanishing altogether in time Three.

We are not demigods and we are not cattle. We are more than our brains but not in the end, I feel, more than the consciousness those brains exist to serve. There is in me something greater and more enduring than anything in my time One experience. But outside or beyond that experience, not in time One, is something infinitely greater and more enduring than anything I can claim as mine. This I realized in that dream or vision of the birds, which I described 25 years ago in *Rain Upon Godshill*, from which I shall quote it. The setting of the dream owed much to the fact that not long before, late at night, I had helped with some bird-ringing at the St. Catherine's lighthouse in the Isle of Wight:

I dreamt I was standing at the top of a very high tower, alone, looking down upon myriads of birds all flying in one direction; every kind of bird was there, all the birds in the world. It was a noble sight, this vast aerial river of birds. But now in some mysterious fashion the gear was changed, and time speeded up, so that I saw generations of birds, watched them break their shells, flutter into life, weaken, falter, and die. Wings grew only to crumble; bodies were sleek and then, in a flash, bled and shrivelled; and death struck everywhere at every second. What was the use of all this blind struggle towards life, this eager trying of wings, all this gigantic meaningless biological effort? As I stared down, seeming to see every creature's ignoble little history almost at a glance, I felt sick at heart. It would be better if not one of them, not one of us all, had been born, if the struggle ceased for ever. I stood on my tower, still alone, desperately unhappy. But now the gear was changed again and time went faster still, and it was rushing by at such a rate, that the birds could not show any movement but were like an enormous plain sown with feathers. But along this plain, flickering through the bodies

Birds in flight in Japan—
a "vast aerial river of birds."

themselves, there now passed a sort of white flame, trembling, dancing, then hurrying on; and as soon as I saw it I knew that this flame was life itself, the very quintessence of being; and then it came to me, in a rocket-burst of ecstasy, that nothing mattered, nothing could ever matter, because nothing else was real, but this quivering and hurrying lambency of being. Birds, men, or creatures not yet shaped and coloured, all were of no account except so far as this flame of life travelled through them. It left nothing to mourn over behind it; what I had thought was tragedy was mere emptiness or a shadow show; for now all real feeling was caught and purified and danced on ecstatically with the white flame of life. I had never felt before such deep happiness as I knew at the end of my dream of the tower and the birds. . . .

And this white flame did not become visible, you may have noticed, until after the second speeding-up of all that bird life, in what could be described as *the third time*.

Readers of Jung will remember the importance he attaches to a process of development he calls "individuation," which, bringing a new relation to the unconscious, transforms the one-sided ego into the broadly-based "Self." (Incidentally, he found a similar process and transformation, expressed symbolically, in ancient Chinese thought.)

Now this must be rarely achieved, and only then, in most instances, toward the end of a longish life. What, then, is the point of it? Why struggle toward a goal overshadowed by the grave? Why at last understand how to live just when you are about to stop living?

Jung neither asked nor answered such questions. But now I believe that his "individuation" and achievement of the "Self" are a preparation for existence outside time One, in times Two and Three. Probably in time Two we move from personality to the essential self, never realized in time One; and that now, in time Two, sooner or later the self must take on, as it were, its final shape and coloring, extending itself to its limits, perhaps those belonging to one of a small number of equally essential types. We must become more completely ourselves before, in our existence only in time Three, finally dissolving into selfless consciousness, as I appeared to do when ecstatically aware only of that white flame.

There is of course purpose in all this. I am not an atheist, but I cannot agree with men who talk about God as if He had once attended a Speech Day at their theological college. That hurrying, trembling, delicate, white flame was not God—but it was numinous. We might say it was moving to and from unimaginable creative Being, both away from and toward a blinding Absolute, possibly through the history of a thousand million planets. For whatever else this universe might be, it is obviously very large and extremely complicated. It must therefore contain innumerable levels of being, about which we know no more than a beetle does about the pro-ceedings of the British Association. We men on earth are probably on a very low level, but we have our task like other and higher orders of beings. As far as I can see—and I claim no prophetic insight—that task is to bring consciousness to the life of earth—or, as Jung wrote in his old age, "to kindle a light in the darkness of mere being."

We cannot perform this service, just as we cannot even enjoy a good life, unless our minds and personalities are free to develop in their own fashion, outside the iron molds of totalitarian states and systems, narrow and authoritarian churches, and equally narrow and dogmatic scientific-positivist opinion. It happens that all three are bearing down on us now, so that while men have always lived in jeopardy, our present position is unusually precarious. We have now arrived at a complicated crossroads, where every turning but one—and that one the least obvious, no great roaring highway—will be disastrous.

I have written this book in the belief that the choice of the right turning, a decision that may be final if not for our whole species then at least for our civilization, cannot be separated from the relations between Man and Time.

New grass rises from around the base of a dead tree. Life, death, renewal—the process continues and endures far outside the straitjackets of false ideas of Time.

Index

Figures in *italics* refer to captions

Illustration credits

Key to picture position
(T) top (C) center (B) bottom (L) left
(R) right and combinations

77 (BR) British Museum: photo John Freeman
78 (TR) Mansell Collection
(B) *Pompeii and Herculaneum* by Marcel Brion, Elek Books Ltd., London
79 (BR) Conzett & Huber, Zurich
80 H. E. Edgerton, M.I.T., Cambridge, Mass.
81 Conzett & Huber, Zurich: photo Paul Senn
83 Erich Lessing—Magnum
84 *The Living Brain* by W. Grey Walter, Gerald Duckworth & Co. Ltd., London
85 Musée Condé, Chantilly: photo Giraudon
86 (TR) Reproduced by permission of the Master and Fellows of Trinity College, Cambridge
87 (T) City of Derby Art Gallery: photo Derby Photo Services Ltd.
88 (CR) *The Splendour of the Heavens* ed. Walter Hutchinson, Hutchinson & Co. Ltd., London
89 (TR) (B) Brown Brothers, New York
91 Courtesy James A. Coleman
92 Ernst Hass—Magnum
96 (TL) (R) *Einstein—A Pictorial Biography* by William Cahn, Citadel Press, New York
(BL) Courtesy Mrs. Sidney Strube
102 Drawing by Janssen, S.E.D.E., Paris
103 Courtesy Professor Fred Hoyle
108 Salvat Editores, Barcelona
109 (TL) Warner Brothers
(TR) Charles Hamblett
(B) Photo Ken Coton
111 (TL) Mansell Collection
(B) British Museum: photo John Freeman
112 (T) Bibliothèque Nationale, Paris
(B) *Great Expectations*, director David Lean, England, 1948
113 (BR) Courtesy Madame Mantes-Proust
(BL) Historisches Museum, der Stadt, Wien
(BR) Photo Gisele Freund—John Hillelson Agency
114 By permission of Jesus College, Cambridge: photo Stearn & Co.
115 British Museum: photo John Freeman
116 Photo Zoë Dominic
117 (BL) (BR) John Wood Collection
118 Photo Houston Rogers
119 Photo Étienne Weill
120 Photo Houston Rogers
122 (TR) John Wood Collection
122–3 (B) Photos Ken Coton
123 (TL) Photo Gisele Freund—John Hillelson Agency
(TR) *Observations* by Max Beerbohm, William Heinemann Ltd., London
125 (TL) British Museum
(BR) Anthony Michaelis Collection
126 (TL) Photo Morris Newcombe

(BL) Society for Cultural Relations with the U.S.S.R.
(BR) From the frontispiece of *Fahrenheit 451* by Ray Bradbury, Rupert Hart-Davies Ltd., London
127 (TL) Mander and Mitchenson Theatre Collection
(BL) Anthony Michaelis Collection
(BR) *Metropolis*, director Fritz Lang, Germany, 1925
129 (TL) S. A. Philips Éclairage et Radio
(BL) Cartoon by Chas. Addams © 1946 The New Yorker Magazine Inc.
130 *Secrets of the Soul*, director Paul Pabst, Germany, 1926
131 British Museum: photo John Freeman
132 (BL) Library of Congress, Washington
(BR) Edwin Smith Collection
133 Theatre World Library
134 (BL) (BR) Mander and Mitchenson Theatre Collection
135 (TL) (BL) (R) Theatre World Library
137 Photo Axel Poignant
139 (TC) Adprint Library
(TR) © Laurence Scarfe
(CR) Courtesy Rupert Davies
(BL) *The Australian Aborigines* by A. P. Elkin, Angus & Robertson Ltd., London © 1954 and 1960 by the American Museum of Natural History; reprinted by permission of Doubleday & Company, Inc., New York
(BR) Photo Axel Poignant
140 (TL) (BL) Photos Axel Poignant
142 Musée Guimet, Paris: photo Michaelides
145 British Museum: photo John Freeman
146 (R) (B) British Museum: photos John Freeman
147 Uni-Dia-Verlag, Grosshesselohe, Germany
151 Photo David Holden
155 Detroit Institute of Arts
157 Mansell Collection
159 Paul Popper Ltd.
161 Carnegie Institution of Washington
163 British Museum
177 Conzett & Huber, Zurich
179 (BL) Photo Mike Busselle
(BC) (BR) *La Donne del Mondo*, directors Gualtiero Jacopetti, Paolo Cavara, and Franco Prosperi, Italy, 1962
182 Photo Philippe Halsman
183 (TL) Photo Sandra Lousada
(TR) Keystone
184 (BL) Photo Henri Cartier-Bresson—Magnum
(BR) U.S.I.S.
186 Photo Mike Busselle
187 Photo montage Mike Busselle
189 © Jules Feiffer

Artists' credits

Model by Barry Learoyd 122/123 ; Brian Lee 228 ; John Messenger 238, 239, 250, 251 ; David Parry 32 (TR) 33 (CR) ; Model by Bob Walden 28 (TL) (TR) ; Leonard Whiteman 90 (TL) (TR) (CR) (BL) ; Sidney Woods 23 (BL)

House credits

Editorial assistants : Joanna Evans, Ken Coton
Art assistants : Gilbert Doel, Roger Hyde, Michael Lloyd, Brigid Segrave
Picture research : Judith Bronowski, Patricia Quick